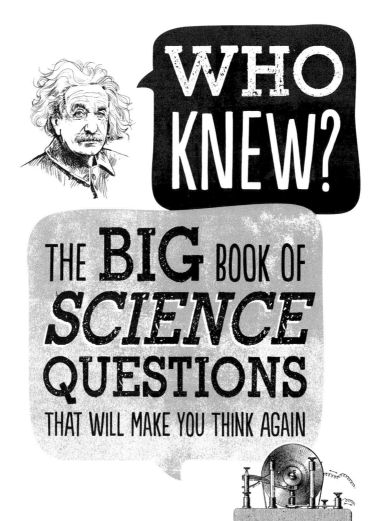

WHO KNEW?

THE BIG BOOK OF SCIENCE QUESTIONS

THAT WILL MAKE YOU THINK AGAIN

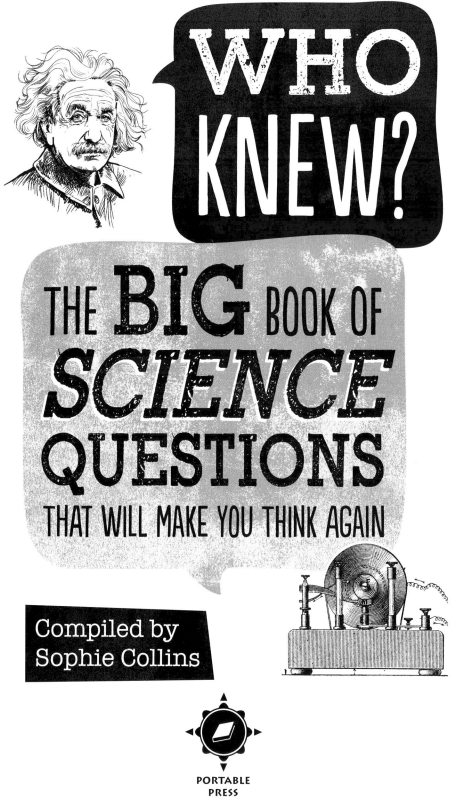

WHO KNEW?

THE BIG BOOK OF SCIENCE QUESTIONS

THAT WILL MAKE YOU THINK AGAIN

Compiled by
Sophie Collins

PORTABLE
PRESS

San Diego, California

Portable Press
An imprint of Printers Row Publishing Group
9717 Pacific Heights Blvd, San Diego, CA 92121
www.portablepress.com • mail@portablepress.com

Copyright © 2022 Quarto Publishing plc

All rights reserved. No part of this publication may be reproduced, distributed, or transmitted in any form or by any means, including photocopying, recording, or other electronic or mechanical methods, without the prior written permission of the publisher, except in the case of brief quotations embodied in critical reviews and certain other noncommercial uses permitted by copyright law.

Printers Row Publishing Group is a division of Readerlink Distribution Services, LLC. Portable Press is a registered trademark of Readerlink Distribution Services, LLC.

This edition contains previously published text, © 2019 Quarto Publishing plc.

Correspondence regarding the content of this book should be sent to Portable Press, Editorial Department, at the above address. Author and illustration inquiries should be addressed to The Bright Press, an imprint of The Quarto Group.

Portable Press
Publisher: Peter Norton
Associate Publisher: Ana Parker
Editor: Dan Mansfield
Acquisitions Editor: Kathryn Chipinka Dalby

Conceived and designed by The Bright Press, part of The Quarto Group,
The Old Brewery, 6 Blundell Street, London, N7 9BH

Design: Lindsey Johns
Cover design: Emily Nazer and Lindsey Johns
Text: Sophie Collins, James Lees and Sarah Herman

Library of Congress Cataloging-in-Publication Data

Names: Collins, Sophie, author.
Title: Who knew? : the big book of science questions that will make you think again / Sophie Collins.
Description: San Diego, California : Portable Press, [2022] | Includes index. | Summary: "This book will not only answer plenty of the questions that you knew you had but will also open your eyes to lots of things that you've probably never even thought about. It's an absorbing read that ranges wide-twelve chapters deal with both the very large (cosmology) and the very small (viruses). Each chapter consists of succinct question-led entries, along with a quiz and some speedy standalone facts for instant "Who knew?" reactions"-- Provided by publisher.
Identifiers: LCCN 2021054027 | ISBN 9781667200743 (hardcover)
Subjects: LCSH: Science--Miscellanea.
Classification: LCC Q173 .C63 2022 | DDC 500--dc23/eng/20220104
LC record available at https://lccn.loc.gov/2021054027

Printed in China

26 25 24 23 22   1 2 3 4 5

"For Ian, for always keeping me guessing!"

CONTENTS

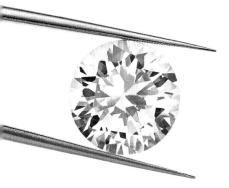

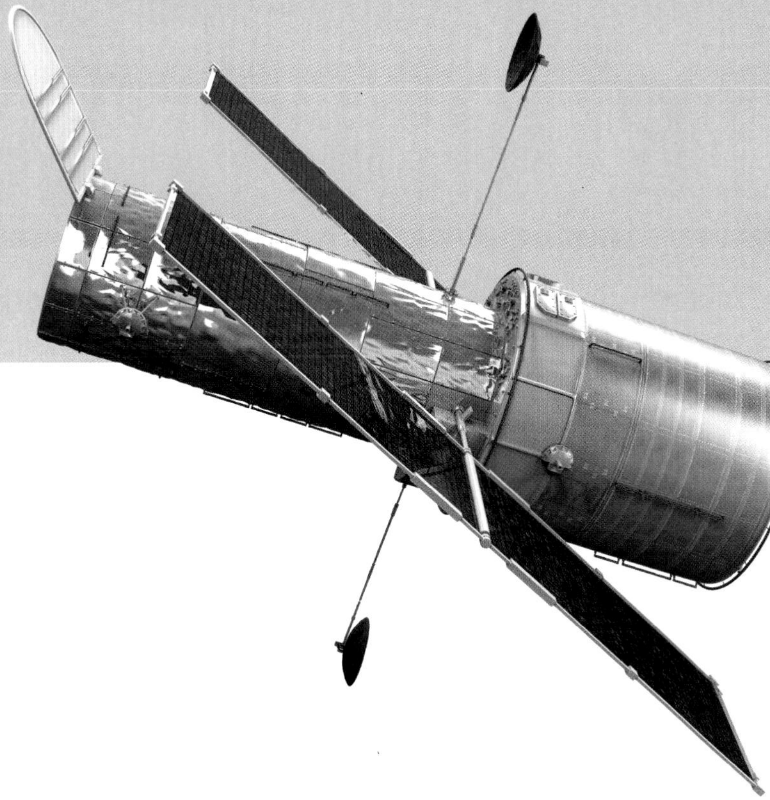

INTRODUCTION

However random life sometimes seems, it's operated by rules, and those rules are dictated by science. Science is behind everything, from the birth of our galaxy to the frankly unsettling dining habits of frogs. It's the reason for qubits, tardigrades, spectral cloaking, and much more (and if any of those sent you on an early visit to the index, your need for this book is even greater than you thought).

Whether you love science already or you've so far been a reluctant guest to the science party, this book will not only answer plenty of the questions that you knew you had but will also open your eyes to lots of things that you've probably never even thought about. It's an absorbing read that ranges wide—twelve chapters deal with both the very large (cosmology) and the very small (viruses). Each chapter consists of succinct question-led entries, along with a quiz and some speedy standalone facts for instant "Who Knew?" reactions. It all makes for an effortless way to explore new terrain, and if you'd secretly like to become a bit more of a science show-off, you'll find plenty of ammunition for that here, too.

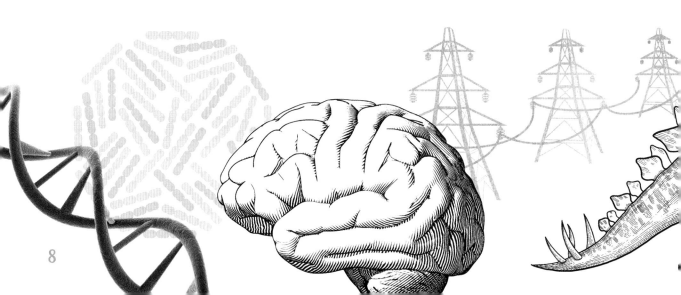

Maybe you'd like to know why planets aren't square, or to be able to name the most poisonous plant in the world? Maybe you've always liked obscure laws and rules—in which case, here's your chance to get up close and personal with the great unconformity, the Fujiwhara effect, the Goldilocks zone, and the Luddite fallacy, to name just a handful. If you like animals, find out why the narwhal takes the ocean's Olympic gold for its echolocation skills; if you're interested in new inventions, discover how close we are to inventing a real-life invisibility cloak. And there are also plenty of opportunities to discover brand-new angles to familiar areas: chapters on the human brain and body offer a quirky owner's manual with excursions into everything from trepanning to rare blood groups, while sections on computing and technology neatly explain the terms we use all the time, from Wi-Fi to fiber optics, often without completely understanding them.

Ready for more? Give your synapses some enjoyable exercise by reading on.

COSMOLOGY

1 AT WHAT POINT DOES THE SKY BECOME SPACE?

Above your head right now is the sky, full of clouds, airplanes, and birds. Somewhere above that is space, with its stars, planets, and galaxies. What you might not have ever considered is exactly where one becomes the other. While there's no definitive answer, the point at which the sky becomes space is generally accepted to be the Kármán line.

THE KÁRMÁN LINE

The Kármán line is defined as sitting at 62 miles (scientists use a figure of 100 kilometers) above sea level. It's based on a rough approximation of the point at which the air becomes too thin to sustain normal airplane flight through lift. The idea is that anything that could fly beyond this point could also fly in deep space, making it a spacecraft. This definition is not exact. Due to the mixing of gases, air currents, and a whole host of other factors, the actual point where conventional flight becomes impossible would likely be different in different places and change over time—so 62 miles is just a rough estimate. It is also unlikely that any actual airplanes would be able to reach such a height, as they are not designed to operate on the very edge of the theoretical flight limit.

The Kármán line is the value set by the French Fédération Aéronautique Internationale, which is the organization that sets the international standards for many things related to aeronautics. This is not to say it's used everywhere. NASA used to use the Kármán line, but reduced its definition of the start of space to just 50 miles to be in line with the U.S. Air Force.

WHAT'S IN THE ATMOSPHERE?

The sky isn't simply one great big single entity. It's a complex system generally defined by several layers that behave a little differently to each other:

The TROPOSPHERE extends up to a round 11 miles into the sky (but lower around the poles). This is the densest part of the atmosphere, and so it has the most going on. About 75 percent of all of the stuff in the atmosphere is concentrated here, and it's the region where most clouds form and airplanes fly.

Above this is the STRATOSPHERE, which reaches up to about 35 miles above sea level. This is where the ozone layer—which protects us from most of the Sun's harmful ultraviolet rays—sits, and it's also where the very highest clouds form. It has also been discovered that some birds are able to fly this high up.

Above this is the MESOSPHERE. Here the air is very thin but just dense enough to be able to cause high-speed objects from space like meteors (see page 39) or space junk to burn up.

Finally, we reach the THERMOSPHERE, which extends to nearly 450 miles. This is where we find the Kármán line, the auroras, and an incredibly sparse collection of atoms that loosely make up the remains of the atmosphere. Beyond this there is only space.

2 WHY ARE GALAXIES FLAT?

Galaxies are enormous clusters of billions of stars. They are truly awe-inspiring to look at (though it can require powerful telescopes). There is a huge amount of variety out there, but they all have something in common. All galaxies are flat, and they are all flat because they all spin.

WHAT EXACTLY IS A GALAXY?

Galaxies are systems of billions of stars, trillions of planets, thousands of black holes, neutron stars, pulsars, and much more besides. They're all held together by gravity. There could be anywhere between 200 billion and 2 trillion galaxies in the universe, and while they vary in size, they can be up to 300,000 light-years (1.7 quintillion miles) across. Pretty much everything in the universe exists inside galaxies. We might think of the space between the planets or even stars as empty, but compared to outside a galaxy, it's practically stuffed to the brim with matter.

GALAXY FORMATION

Galaxies form in the same way as everything else in space—gravity pulls things together. The first galaxies would have been truly titanic gas clouds many times the size of the galaxies we see today. These condensed, forming stars, planets, and everything else. As they did so, all of these newly created stars were

also pulled together by gravity. This process pulls a lot of stuff into the middle of the growing formation, which creates a galactic center where most of the stars are concentrated, making it appear as a big bright bulge. Much of the rest of the material then becomes spread out around the center in a wide rotating disk. There is also a "halo" of disparate stars and other stuff that surrounds the galaxy in a light, spherical cloud of matter.

WHIRLING SILENTLY IN SPACE

As a galaxy is forming and pulling in matter, it starts to spin. In the same way as when a star or a planet forms (see page 14), the act of pulling something toward it starts a rotation, which then makes other things nearby begin to rotate in the same way, until you have a giant ball of matter that is spinning as it condenses down. Because of this spin, the centripetal forces push outward along the area perpendicular to the motion of spin. This means that while most of the stars and other matter are pulled inward along a particular axis, they're pushed out and flattened—like a chef spinning pizza dough in the air. Galaxies are almost like ringed planets, only where the rings are bigger than the central part, and rather than just a bit of rock and ice they are instead made of billions and billions of stars.

3 WHAT ARE BLACK HOLES?

You've heard about black holes; they're dangerous, exciting, and open the door to a million adventures or future technologies. But what are they really? This is not an easy question to answer, and they're still far from fully understood, but black holes are basically massive objects that have such great gravitational pull that nothing can ever escape.

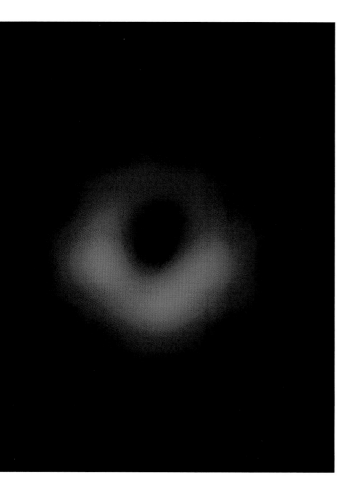

HOW TO MAKE A BLACK HOLE

If you take some matter and let gravity pull it together, you might get something like Jupiter—just a big ball of gas. If you have even more matter, you get a star. If you let that star undergo various processes, it may end up as a very dense object like a neutron star (a thimbleful of neutron star weighs more than Mount Everest) or a white dwarf, where it's full up to the point where the strongest atomic forces pushing out are only just in balance with the gravity pulling in. To make a black hole, you just add some more mass. When you reach a certain amount of mass in a small enough space, then physics starts to get really weird. With enough mass crushing in under gravity, it is eventually enough to break down and cause all of the mass to become a single point. Just an infinitesimally small dot that contains an enormous gravitational pull. This is a black hole.

THE EVENT HORIZON

Gravity pulls us toward massive objects, and black holes are about as massive as things get. They have so much gravity that it can become impossible to escape. While the black hole itself is only a small dot, it has around it an "event horizon"; this is the point at which its gravity becomes so strong that even light can't get away. The event horizon means that when we look at a black hole we see a dark, empty circle in space, as it eats anything that might shine there—hence the name of "black hole."

BEYOND THE HORIZON

Let's say you were in a spaceship that could travel faster than the speed of light, so you decide to nip into an event horizon and see what is going on. This is what you would find. As you cross over the event horizon, nothing. There is no change that you can see or even initially detect. Bored and likely disappointed, you turn your ship around and power up the faster-than-light engines and blast away. Or rather, you blast deeper into the black hole! The black hole's gravity is so strong that it actually causes space to fold in on itself, making all directions of space point toward its center. You are now trapped, and as you get closer to the black hole itself the gravity at your legs is stronger than at your head, as your legs are closer to the center, so you start to get stretched out in a process called spaghettification. This will all take some time. Albert Einstein showed that space and time are really the same thing, and because space is messed up, so too is the time. In fact, as you get closer to the center of the black hole, time will start to slow down more and more. As you look out into the universe you will see stars being born and dying before your very eyes as the star's time moves faster compared to you, until eventually at the black hole itself time may just stop. (We're not sure on this; physics gets really messy in a black hole, so it's a best-guess scenario.)

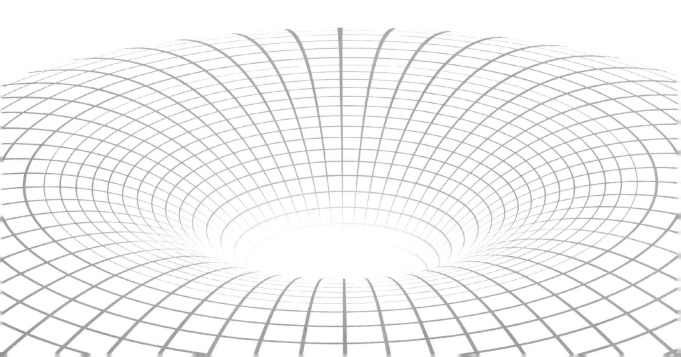

5 WHERE ARE THE STRONGEST TELESCOPES ON EARTH?

For a telescope to work effectively on Earth, it needs to be sited somewhere high, dry, and preferably remote. The less light pollution there is, and the clearer the air, the better the quality of its images.

Light pollution has actually seen some observatories moved altogether. Mount Wilson was the foremost U.S. site until the light from nearby Los Angeles made it increasingly ineffective, and it was moved to Mount Palomar in the San Diego mountains. Another factor that may affect results on the ground is weather. While Las Campanas Observatory, situated high on a peak in Chile, can enjoy as many as 300 good viewing nights in a year, Mauna Kea in Hawaii, otherwise an extremely successful site, loses far more due to adverse weather.

JUDGE BY THE RESULTS

A good viewing result depends on what you're looking for. The Gran Telescopio Canarias in the Canary Islands looks into deep space. In 2016, it captured an image of stars around a galaxy located 500 million light-years from Earth—going ten times deeper than any other taken from the ground. If you want to look at pictures of the Sun on the other hand, go for the Inouye Solar Telescope in Hawaii. Images of the Sun's surface taken by the Inouye show thousands of small areas

separated by darker divisions—all the more striking when you grasp the scale of what you are looking at: each tiny "cell" is around 200,000 square miles. And if you want to boost your knowledge of dark energy which, despite the fact that no one can define it, is estimated to take up anything from 70 to 85 percent of the universe, go for the duo of Keck telescopes on Mauna Kea.

NEW SPACE STAR

Of course there's no rule that telescopes stay earthbound. NASA's Hubble Space Telescope, launched back in 1990 and is still in operation, was perhaps the first celebrity telescope—a household name, with space discoveries that were regularly televised. But the James Webb Space Telescope, also from NASA—launched in December 2021 after a series of delays and at a cost of $10 billion—is a hundred times more powerful than the Hubble and should be able to look at objects so remote that the view of them is already 100 million years in the past.

PROS AND CONS

If space telescopes can offer such awe-inspiring results, why continue to build the Earth versions? Well, they're cheaper, more accessible, and easier to maintain. And there's also no risk that billions of dollars' worth of technology will be damaged or destroyed by floating space junk.

Equipped with a sun shield to protect it from the searing heat, it will be sent into an orbit of the Sun, 930,000 miles from Earth. With an 18-part mirror that is 21 feet 4 inches wide, featuring gold-plated hexagonal panes of beryllium, an unusually light metal, it seems destined to hold the crown as the most powerful contemporary telescope, at least for a while.

6 WHAT'S THE HOTTEST PLACE IN THE UNIVERSE?

There is a theoretical maximum temperature known as the Planck temperature, which in Fahrenheit is at 2.5 with 32 zeros after it; however, nothing has ever come close to this, and it is unlikely anything ever will. The hottest place in the universe is inside the Large Hadron Collider (LHC) at the CERN facility on the border of France and Switzerland.

THE HOTTEST THING

The LHC works by accelerating particles up to near the speed of light and crashing them into each other. By smashing two gold particles into each other, it is possible, if only for a fraction of a second, to reach temperatures of 7,200,000,000,000°F. This is twice the Hagedorn temperature, which is the point where normal matter breaks down. Scientists exploit this fact and use it to split atoms apart into various components, which they can then examine within the LHC.

THE HOTTEST NATURAL THING

There are a lot of hot things out there in space. Our own Sun can reach several millions of degrees (see page 36) and even larger stars can get nearly ten times as hot. There is also a cluster of galaxies called RXJ1347 which are colliding and have heated up to over 540,000,000°F. The hottest thing, however, is in the core of a supernova. Supernovas are enormous explosions caused by collapsing stars. The enormous pressure within a supernova is able to produce temperatures in excess of hundreds of billions of degrees.

7 WHAT'S THE COLDEST PLACE IN THE UNIVERSE?

The coldest anything can get is 0 degrees kelvin (-459.67°F). This is the point where everything stops moving, even atoms. Quantum physics tells us that it is not possible to reach this temperature, so what is the closest thing? The coldest temperature ever recorded was on a tiny piece of metal in Colorado.

THE COLDEST THING

In 2016 researchers at the National Institute of Standards and Technology in Boulder, Colorado, took a tiny piece of aluminum (only 0.02 millimeters wide) and, using a special laser technique called "sideband cooling," managed to cool it down to just 0.00036 degrees kelvin.

Low temperatures make it a lot easier to study much of physics. Temperature is a measurement of the movement of atoms within an object. At lower temperatures everything is moving around a lot less and unexpected things are less likely to happen, so there are a lot of reasons for continuing this kind of work. Research teams in Gran Sasso, Italy, and Lancaster, England, are among those who have previously held the record for coldest place in the universe and scientists may, in the future, be able to create even colder temperatures still.

THE COLDEST NATURAL THING

The coldest naturally occurring thing in space is the Boomerang Nebula. It's a star at the end of its life that is throwing out a huge amount of frigidly cold gas. This gas is being forced out at such a rate that it expands rapidly, causing it to become supercooled. This lowers the temperature of the space around it down to just 1 degree kelvin—colder even than empty space, which sits at 2.7 degrees kelvin.

8 WHAT MAKES DARK MATTER DARK?

You may have heard about dark matter. It's one of the greatest mysteries in our universe, and even its name alludes to the things still to be learned about it. Dark matter is so called because it doesn't give off any light, and that's about all we know about it.

SOMETHING OUT THERE

There is a lot of dark matter out there. In fact, there is more than five times as much dark matter as there is normal matter. But if we've never seen it, and we don't know what it is, how do we know that there is any dark matter at all? Let alone how much of it there is.

In 1933 a scientist named Fritz Zwicky was working on calculations of a galactic cluster millions of light-years away. He determined the mass based on how bright it was and then did his calculations. He quickly realized that they were moving faster than expected. The only explanation was that there was missing matter not contained within the brightness, some forty times more than what could be seen. He therefore dubbed this "dark matter." It took a while, but it became clear that it wasn't just this distant galactic cluster that was missing matter. Every cluster was missing mass, and it turned out that even our own galaxy

DARK ENERGY

Dark matter isn't all that's missing from our understanding of how the universe works. Calculations on the expansion of the universe throw up the question of unknown energy that seems to fill empty space. This has been called dark energy for its mysterious origin and makes up nearly 70 percent of everything in the universe. Scientists know even less about this than they do about dark matter.

was, too. The solution to all of these problems was dark mass—some invisible form of mass that sits on the edge of galaxies and galaxy clusters, forming a huge web throughout the universe.

MYSTERIOUS MATTER

We don't know what dark matter is, but there are many theories as to what it could be. An initial suggestion was that there are a lot more black holes, planets, and brown dwarf stars out there than we thought. These are essentially invisible to modern technology and provide some mass. However, even the wildest estimates don't get close to covering the missing mass. The leading theory for what dark matter might be is weakly interacting massive particles (WIMPs). These would be tiny particles that interact with the universe only through gravity and the weak nuclear force (the weakest of the four fundamental forces), making them even harder to detect than neutrinos. Some of the more imaginative theories suggest that perhaps gravity doesn't work in the way we think it does on such large scales, or that perhaps gravity is somehow coming from other dimensions. While these ideas should not be discounted, there is little compelling evidence to believe that they are possible.

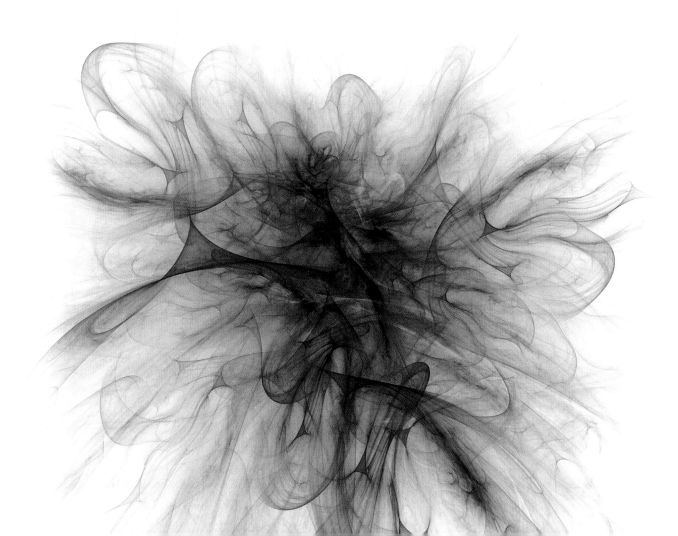

9 HOW BIG IS THE UNIVERSE?

As Douglas Adams once famously wrote, space is big. Really big. The observable universe is a whopping 46 billion light-years across, but the universe itself could be much bigger.

UNIVERSAL EXPANSION

Ever since the big bang, the universe has been getting bigger. The big bang may even be more accurately called the big expansion. The universe went from an infinitesimally small dot and got bigger very quickly. In the inflationary epoch, it went from about 1 nanometer across to nearly 11 light-years in less than a hundred billion billion trillionths of a second (1 with 32 zeros in front of it). After this rapid expansion, it slowed down dramatically but continued getting bigger at something like the speed of light. The universe is about 13.8 billion years old, so it would make sense that the edge of the universe is 13.8 billion light-years (plus that initial expansion of 11 light-years) from the middle, and then if that goes both ways, the total width of the universe is about 27.6 billion light-years across. But it's not quite that simple.

IT KEEPS GETTING FASTER

In 1998 two teams studying distant supernovas realized that not only is the universe expanding but the rate of that expansion is increasing. It also became obvious that this was a relative effect: Everything is moving away from everything else, and the farther apart objects are from one another, the faster they move away. This places a limit on how far we can see. Objects far enough away from us could move faster than the speed of light farther away from us, so we can never see them. Calculations cap the farthest we can see at a universal width of 46 billion light-years. The universe, though, likely continues beyond this and could, in theory, be infinitely large.

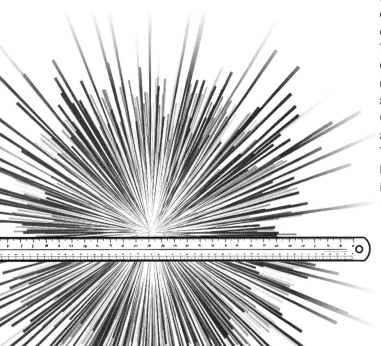

QUIZ
COSMOLOGY

Reach for the stars and see what you've learned in this chapter—you'll have a blast testing your friends on their knowledge of the universe!

QUESTIONS:

1. How high does the troposphere reach above Earth's surface?

2. How do galaxies form in space?

3. Where are you able to see the clearest pictures of the Sun from Earth?

4. What happens to time in a black hole?

5. How old is the universe?

6. What's the coldest thing recorded on Earth?

7. Who invented the term "dark matter"?

8. How big is a galaxy?

9. Why is Mount Wilson Observatory no longer the foremost site for looking into space in the United States?

10. What is spaghettification?

Turn to page 244 for the answers.

ASTRONOMY

10 HOW DO SOLAR ECLIPSES OCCUR?

Solar eclipses are truly magnificent events—the sky falls dark during the day and a fantastic halo of light surrounds the Moon. They happen when the Sun, Moon, and Earth line up perfectly in an event called a "syzygy." The Moon moves into place between the Earth and Sun, blocking out the Sun's light and causing an eclipse.

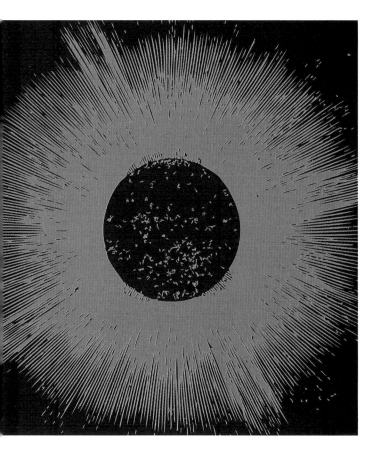

UNIQUE IN THE UNIVERSE?

If the Earth ever were to become part of some great intergalactic alien civilization, then it would probably be a tourist hotspot for its eclipses! The alignment of three objects in space is not rare, but a total solar eclipse certainly is. While there is some small variation due to orbital patterns, the Moon and the Sun look almost the same size in the sky when viewed from Earth. This is an incredible coincidence not seen anywhere else in our solar system, which leads to the striking ring of light seen during an eclipse. Astronomers believe it to be very rare and possibly even unique in the universe.

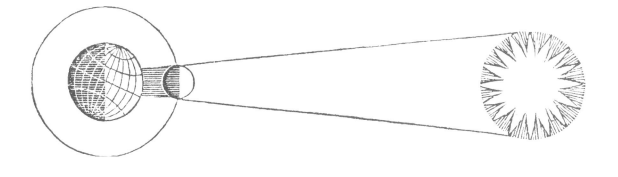

A SIGN FROM THE HEAVENS

Eclipses are difficult to miss, and they have been noted by humans ever since records began. In the past they were often seen as omens of great importance; many civilizations have tried to link the deaths and births of great historical and religious leaders or major events to eclipses. The ancient Greeks in particular put great stock in eclipses. The occurrence of an eclipse during the Battle of Halys in 585 BCE reportedly caused the troops to stop fighting, resulting in the swift arrangement of a peace agreement. This one had been predicted beforehand by Thales of Miletus, suggesting that the ancient Greeks at least partially understood how and why eclipses occurred.

OTHER TYPES OF ECLIPSES

If the Moon can get in the way of the Sun, then it makes sense that the Earth can get in its way, too. When this happens, the event is called a lunar eclipse. A lunar eclipse is often better known as a "blood moon," because as the Earth passes between the Moon and the Sun, only the red light from the Sun is able to reach the Moon's surface, making it briefly appear a deep, rich red color.

THE MOON AND THE EARTH

The orbit of the Moon around the Earth is not on a flat plane with the Earth's orbit around the Sun. This is why we don't get a solar eclipse every month. This wonkiness in the orbits means that we can sometimes get partial eclipses, where only part of the Moon passes between the Earth and the Sun. It's not only the Moon that does this. The planets Venus, Mars, and Mercury all at some point in their orbit pass between the Earth and the Sun. However, because the planets are much farther away from Earth than the Moon is, they aren't able to cause the same effect, as they appear very small. When a planet passes in front of the Sun, it is known as a transit—it is possible to see one using a specially equipped telescope.

11 WHY ARE THERE NO SQUARE PLANETS?

Planets come in all sorts of sizes but only ever one shape: round. A planet can be made of rock, ice, or even gas, but the shape is always the same. This is because of the planet's own gravity pulling inward.

IT'S ALL IN THE MAKING

Gravity pulls everything toward everything else. The bigger the thing, the more pulling power it has. Gravity will always pull things toward the center of an object. This means that as a big object starts to form, it is going to have a very strong pull all over and will try to get as much stuff as close to the middle as possible—and the best shape for this is a ball. If a planet were a cube, then the corners would be farther away from the center than its faces, so the gravity would naturally pull on them until they flattened out, forming a sphere.

While this might make sense intuitively for gas planets, which are able to change shape easily, what about the more solid rock and ice planets? Their spherical shape is set when they are being formed. They start life as many small rocks or chunks of ice that are pulled together by gravity in the same way that clouds of gas are pulled together during the formation of a gas planet. However, as new solid material is pulled into the forming planet, this causes huge collisions. These crashes release a large amount of heat and the planets are turned into molten rock (in the case of rock planets) or liquid (in the case of ice planets) that gravity is then able to pull into the round shape we see today.

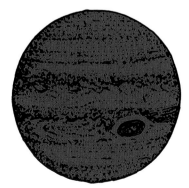

WHAT'S NOT ROUND?

It's not just planets. Stars, black holes, and many other things in the universe are also made round by gravity. So what isn't? The short answer is anything that is smaller than about 370 miles wide, as it won't produce enough gravity to make itself spherical. This means that things like asteroids and comets can be all sorts of shapes. They are often roughly spherical, but they can come as long cylinders, strange lumpy blocks, or even (in the case of the comet 67P/Churyumov-Gerasimenko) sort of duck-shaped.

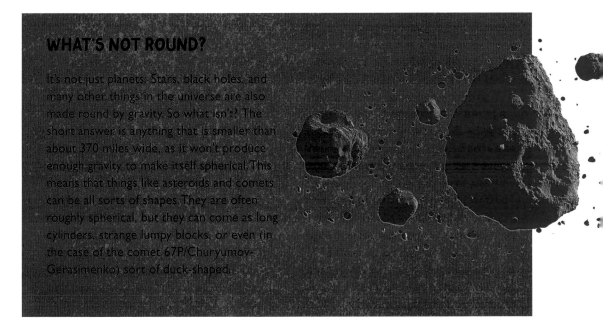

IS THE EARTH REALLY ROUND?

The Earth, like many planets, is not perfectly round. Its rotation and gravitational forces cause it to squish a little at the poles and bulge at the equator. The difference is about 17 miles between the Earth's width at the poles and the equator. The surface of the Earth is also not flat: The deepest point is the Mariana Trench at a little under 6.8 miles below sea level, and the highest point is the peak of Mount Everest at just under 5.6 miles above sea level.

The Earth is on average a whopping 7,918 miles wide (gravitational forces mean it varies by up to 13 miles). So what if we shrunk it down to the size of a billiard ball at just $2^1/4$ inches wide? The difference between the width at the poles and the width at the equator becomes just 0.0078 inches and the difference between the highest and lowest point on Earth a mere 0.0039 inches. Official billiard balls have a roundness tolerance (as defined by the World Pool Billiard Association) of about 0.0051 inches, meaning that the Earth is not quite as round as a billiard ball.

12 WHAT ARE COMETS MADE OF?

With their fantastic streaking tails, it's no wonder that comets capture the imagination of young and old alike. But what are these infrequent visitors to the night sky actually made of?

BIG, DIRTY SNOWBALL

Comets are made of a mixture of rock, dust, ice, and frozen gases—a bit like a big, dirty snowball. The main body of a comet can be anything between a few hundred yards and several miles wide. The central part, called the nucleus, is made of a dusty rock surface covered in ice and frozen gases such as carbon dioxide, methane, and ammonia. Comets may also contain substances with more complex molecules, such as formaldehyde, ethanol, and potentially even hydrocarbons and amino acids. These are the chemicals that form the basis of life, and some theories suggest that perhaps it was comets crashing into the Earth during its formation that allowed life to start here.

STREAKING TAILS

The rather striking tail that can be seen trailing behind a comet is caused by the frozen gases on the surface of the comet melting in the heat of the Sun and being ejected off into space. Comets begin to form their tails when they get about 370,000 miles away from the Sun. Some of these tails can stretch out for millions of miles. It is worth noting that because a comet's tail is formed by the Sun and its solar winds, a comet's gas tail will always point away from the Sun, regardless of its direction of travel.

THE MOST FAMOUS COMET

Probably the most famous comet in the world is Halley's Comet. Its orbit around the Sun means that it comes into view of the Earth every 75 to 76 years. It has been recorded in history since antiquity, but it was only in 1705 that the English astronomer Edmond Halley realized that it was the same object that kept appearing in the sky and predicted when it would next appear (which was unfortunately 17 years after his death). The first record of Halley's Comet was in 240 BCE from ancient China, and it was also recorded by the Babylonians in 164 BCE and 87 BCE. Perhaps the most famous recording of the comet is in the Bayeux Tapestry, where it is displayed as an omen during the Norman conquest of England. The comet will next be visible in 2061.

LANDING ON A COMET

On November 12, 2014, the European Space Agency's Philae module touched down onto the surface of a comet. While it wasn't a perfect landing, with the lander bouncing twice and then coming to a stop in a crack, it was the first of its kind. Despite falling on its side, the lander was able to complete most of its scientific objectives, identifying a number of chemicals that hadn't been detected on comets before. It was also able to measure the temperature shift across the comet's 12.5-hour day from -292°F to -229°F and discover that the comet's surface is a 4- to 20-inch-thick crust of dust held together with ice, with a more "fluffy" porous rock layer beneath.

13 HOW HOT IS THE SUN?

It might sound like a simple question and a quick search on the internet will give you an answer of around 10,000°F, but in truth it's not that easy. The Sun is a massively complex object with many different layers that each have a different temperature.

THE STRUCTURE OF THE SUN

The Sun has a huge central core that acts as a furnace, fusing hydrogen atoms together into helium at a temperature of 28,300,000°F. Around this are several layers of hot, dense plasma that can be up to 12,600,000°F. The surface is a relatively cool 10,000°F; however, extending out beyond this, things can get even hotter! The solar winds and corona, which are caused by the Sun's massive magnetic fields, reach temperatures of up to 9,000,000°F. But even these temperatures pale in comparison with the solar flares, which are enormous jets of solar plasma that can reach 36,000,000°F!

FLYING CLOSE TO THE SUN

The Sun is very hot, so if you tried to fly by in a spaceship you'd get cooked, but what is the closest you can get? A space suit could keep you safe up to about 250°F and a spaceship up to nearly 5,000°F, so in a good spaceship you could get to about 1.3 million miles away from the Sun. The object that humans have put closest to the Sun is the Parker Solar Probe, launched in August 2018, which at its closest approach to the Sun will be 4.3 million miles away.

FIVE QUICK FACTS

 1 TWINKLE, TWINKLE LITTLE STAR

It's the distortion of the light traveling through Earth's atmosphere that produces the sparkling effect; out in space, stars don't twinkle.

 2 PLANET NAMES ARE REGULATED

The International Astronomical Union, founded in 1919, sifts through potential names for newly discovered planets, comets, moons, and asteroids to make sure there aren't any duplicates (names that are judged stupid or offensive are ruled out, too).

 3 ASTRONOMERS HAVE NAMES FOR THINGS THEY HAVEN'T FOUND YET

For example, when (and if) they find a moon that orbits another moon, it will be called a moonmoon. Although no astronomer has seen one, when they do, the name is all ready for it.

 4 COMETS' TAILS FLOW AWAY FROM THE SUN

Comets are made up from a mixture of ice, ammonia, carbon dioxide, and other chilly ingredients, and run in a perpetual orbit of the Sun; as they pass close, some of the frozen material is heated straight to vapor and flows backward off the comet, forming the gaseous "tail."

 5 THE MOON ISN'T ROUND

When it rises in the sky, it looks perfectly circular, but the Moon is actually an oblate spheroid, which means it's closer to the shape of an egg. It appears round because you only see the part of the Moon that's lit by the Sun.

14 WHAT IS A SHOOTING STAR?

A streak of light across the night sky—a shooting star! Maybe you feel like you should make a wish on it? While there's nothing stopping you, you should probably know that a shooting star isn't actually a star—it is caused by a piece of rock or space junk falling through the Earth's atmosphere.

A SHOOTING STAR IS BORN

When rocks or some of the millions of tiny pieces of space junk fall toward Earth, they get pulled in by gravity. They get faster and faster, reaching speeds upward of 23,000 feet per second. When they enter the Earth's atmosphere, they start to collide with air particles and other molecules, which produces friction, causing them to get very hot and burn up, forming large streaks across the sky. This usually happens at around 30–60 miles above the ground and the resulting shooting star lasts only a second or so. Shooting stars can happen all year round but are more common during meteor showers, when the Earth passes through a stream of debris left behind by something such as a comet.

BIGGER AND BIGGER

Most shooting stars that you'll see are caused by something that's between the size of a grain of sand and a pebble, but they can get much bigger than that. Meteorites are the remaining bits of rock that can be found on the ground after a shooting star falls to Earth. To leave behind a rock that would fit comfortably into your hand, the original rock would have to have been about 3 feet or so in size. The largest meteorite ever found is known as "Hoba," after the farm where it was discovered. It weighs a whopping 66 tons and is a square roughly 10 feet wide and 3 feet thick.

METEOR STRIKE

As you can probably imagine, the impact of tons of rock traveling at thousands of feet per second can be incredibly dangerous. Fortunately, the most destructive events are very rare. The most well-known event of recent memory was the Chelyabinsk meteor in 2013, which was a 65-foot asteroid that was traveling at around 12 miles per second. As it passed through the atmosphere it turned into a huge fireball that could be seen over 62 miles away, and people close to it felt the intense heat as it exploded in midair. While spectacular, the event caused only minor damage to buildings.

Some meteors can be far more devastating. The Tunguska event, which took place on June 30, 1908, is thought to have been a meteoroid of between 200 and 620 feet exploding in the atmosphere. It blew up with a force 1,000 times greater than the atomic bomb dropped on Hiroshima and flattened an area of over 1,200 square miles, knocking down close to a hundred million trees. Fortunately, it occurred in a remote area of Siberia, so no people were hurt in the blast. One of the largest meteor impacts in the Earth's history caused the Chicxulub crater in Mexico. The meteoroid, estimated to be between 6 and 9 miles wide, landed 66 million years ago, and it is thought by some that the material thrown up into the atmosphere after the impact caused massive climate disruption, in turn triggering the extinction of the dinosaurs.

15 WHAT HAPPENS WHEN ALL THE PLANETS ALIGN?

It's a staple of Hollywood films, prophecies, and astrology. When the planets in our solar system line up, there will be some sort of event of great power. Even the less fantastical accounts say that there will be strange gravitational effects that could cause chaos. But would it even be possible for all the planets to line up?

A RARE OCCURRENCE

In fact, the planets in our solar system can never truly align. The planets orbit around the Sun at different angles, so it would never be possible for them to form a line. Because the planets orbit at different speeds, it is possible for all the planets to be in roughly the same patch of sky as seen from the Earth, but even at the point at which they are as physically close to each other as possible, they will still be fairly spread out. The next time that all of the planets will be visible in the sky at the same time won't occur until 2492, and even then they will be spread across a large area. Events like that only occur once every several thousand years.

WHAT ABOUT GRAVITY?

Alarmists have suggested that planetary alignment could cause the effects of gravity to multiply, but this is simply untrue. The other planets in our solar system do have a gravitational effect on the Earth. However, this effect is incredibly tiny—so small, in fact, that even if it was somehow possible for them to line up, it wouldn't actually have any effect on our planet.

16 HOW MANY MOONS DOES EARTH HAVE?

Moons are celestial bodies orbiting around a larger body, such as a planet, which is presumably orbiting around a star. Other planets, such as Jupiter, have numerous moons. At the last count, there were sixty-seven of them orbiting the gas giant. That makes our Moon seem rather lonely by comparison, but in fact it isn't alone up there.

TEMPORARILY CAPTURED OBJECTS

The Moon is 2,175 miles across and has been orbiting Earth for over four billion years. However, there are thousands of other temporarily captured objects, or "mini moons," caught in Earth's orbit, some of which measure a few feet across, and most of which are even smaller. In 2006 a sky survey conducted by the University of Arizona found a mini moon the size of a car. It was named 2006 RH120 and left Earth's orbit to resume orbiting the Sun less than a year later. In March 2011, a team of scientists used a supercomputer to calculate that at any given time there is at least one asteroid orbiting the Earth that measures 3 feet across. These asteroids don't orbit the Earth in neat circles—the twisty path they follow is a result of their being pulled between the gravity of Earth, our Moon, and the Sun.

MORE MOONS?

It is possible that in the past Earth had a second large moon. That might explain the strange terrain on the far side of the Moon, which could be the result of our Moon crashing into another. Moons come and go—Mars currently has two large moons, but one of them is headed straight for the planet, where it's expected to crash in the next 10 million years. There's also a chance our planet could acquire a second large moon in the future.

17 WHERE IS THE SUN'S TWIN?

Most Sun-like stars are born in pairs, and the Sun itself is likely no exception. However, when we look into the sky there's clearly only one sun up there. So either there never was a twin, or there was but now nobody knows where it is.

THEY COME AS A PAIR

A large number of all medium to large stars are born in pairs. This is because of the circumstances of their birth. Stars are generally made in enormous clouds of dust and gas that are pulled together by gravity. As the particles are pulled together, they rub against each other. As the stars get larger and larger, they manage to generate enough heat and pressure to ignite. Because this process takes place in truly enormous clouds, more than one star will form at the same time and the act of one star forming may even help other stars to grow.

A STAR CALLED NEMESIS

The long-lost twin of our Sun is often dubbed "Nemesis," after the ancient Greek god of retribution. The name comes from the theory that the separation of the two stars and the twin star's later travels through the solar system caused it to throw huge amounts of comets and asteroids at the Sun. The original theories about the twin star suggested that Nemesis was a red or brown dwarf that still orbits the Sun at the very edge of the solar system. However, modern technology that is capable of surveying the farthest reaches of our solar system has failed to find anything. If there ever was a Nemesis, it has long since left the solar system.

BINARY STAR SYSTEMS

Over half of all the Sun-like stars that astronomers have observed come as a pair in what's called a binary star system. This is where the two stars orbit around some central point between them. These systems can also have planets around them. Systems with three or more stars in them can be very chaotic and prone to collapse or break apart. This is not to say it's not possible, but such systems often take the form of a binary star system with another star a long way out orbiting around both of them.

WHERE CAN MY LITTLE STAR BE?

The question remains: Where, then, is the Sun's twin? It's certainly not within the solar system, or even anywhere nearby. Searches conducted by powerful telescopes are hampered by gas clouds near to Earth that can block our view of regions of space— and despite our best efforts, nothing has been found. What has been found, though, is one of the Sun's other siblings. Identified through the composition of its elements, the star HD 162826 lies in the constellation of Hercules. The star isn't bright enough to be seen with the naked eye, despite being slightly larger than the Sun, but research suggests that it was born in the same place that our own Sun was. So there may yet be hope of finding the wayward twin.

18 WHY DO WE ONLY EVER SEE ONE SIDE OF THE MOON?

The Moon is an ever-present feature of our sky. Even looking at it with the naked eye, we can see that its surface is very distinctive. But if it's an orbiting body, then why does it always look the same? Why do we always see the same part of the Moon? It's because the Moon is tidally locked to the Earth.

HERE'S LOOKING AT YOU

Tidal locking occurs when one object orbits another and, over a sufficiently long period of time, the gravitation tidal force (which is the pull of gravity between the Earth and the Moon) causes the rotation to slow down or speed up. What this means for the Moon is that gravitational pull from the Earth has made the rotation speed of the Moon slow down to a point where it orbits around the Earth every 27.3 days but also takes 27.3 days to spin on its axis. This means that the same side is always facing the Earth. It is not only the Moon that is affected by tidal locking; the locking process significantly helped to slow the rotation of the Earth to the 24 hours it is now from the 6 hours it once was. And the process is still ongoing—a year gets longer by 0.000015 second per year.

A UNIVERSE—WIDE PHENOMENON

Any two orbiting objects in space that are sufficiently large and close to one another will eventually become tidally locked. Usually, the smaller body becomes tidally locked to the larger one. Most of the major moons in the solar system are tidally locked to their planets, and the planet Mercury will likely become tidally locked to the Sun at some point in the future. The dwarf planet Pluto is tidally locked with its moon Charon, which is of a similar size. This means that if you were standing on Pluto's surface, not only would you always see the same side of Charon, but it would also always be in the same place.

THE TWILIGHT ZONE

A planet that is tidally locked to a star would be a strange place indeed. It would have one side facing the star in constant day, which would likely become very hot. The other side would always be facing away and thus be in a permanent cold night. If the planet had enough material on it and some kind of atmosphere, the planet would likely be split into a frigid iced plane covering half of the planet and a baked desert on the other side. The only potentially habitable zone would be a thin band around the equator, where enough of the hot air would be able to keep the ice melted and something of a water cycle would be able to start. Life might just be able to make a home in such an eternal dim twilight.

THE DARK SIDE

Since the same side of the Moon always faces the Earth, it also stands to reason that the same side of the Moon always faces outward. This "dark side" of the Moon is, therefore, less shielded from meteor impacts, and because of this it is heavily cratered.

19 WHY DOESN'T THE NORTH STAR MOVE?

The North Star has been used throughout the years as a clear point of reference for navigation, since it always stays in the same position in the sky, pointing north. The reason that it alone stands sentinel while all the other stars move? The axis of rotation (the imaginary pole around which the Earth rotates) is pointed directly at the North Star.

ON THE MOVE

The Earth's rotational axis doesn't stay perfectly still, it moves slowly over the years. The position that the axis points to moves in a circle over a period of about 26,000 years. While Polaris holds the title of the North Star at the moment, in antiquity a dark spot at the midpoint between the stars Alpha and Beta Ursae Minoras was used for the same purpose. Thousands of years in the future, stars like Vega, Alderamin, and Thuban will all have their turn as the North Star.

THE SOUTH STAR

On the other side of the planet, Sigma Octantis currently holds the title of South Star, but it is very dim and only barely visible on a clear night, so it is functionally useless for navigation. The constellation of the Southern Cross is used instead, as it points roughly to where the South Star would be. As with the North Star, the South Star changes over time as the Earth shifts. In some 60,000 years' time, Sirius, the brightest star in the night sky, will become the South Star.

QUIZ

ASTRONOMY

Were you paying attention, or did the facts whizz past you like Halley's Comet? Try this quiz to see how much you remember about space.

QUESTIONS:

1. What is a syzygy?

2. What's the other term for a blood moon?

3. How much does the width of the Earth vary, and why?

4. What's the highest temperature to which a space suit could protect you?

5. How far from the Sun does a comet have to be before it starts to grow its tail?

6. When was the earliest recorded sighting of Halley's Comet?

7. What's the biggest meteorite ever found on Earth?

8. Will Polaris always be the North Star?

9. When will all the planets next be visible in the sky at the same time?

10. How many moons does Mars have?

Turn to page 244 for the answers.

EARTH MATTERS

20 WHAT'S THE DIFFERENCE BETWEEN A TYPHOON, A CYCLONE, AND A HURRICANE?

Hurricanes, typhoons, and cyclones sometimes seem to be used almost interchangeably when weather is under discussion. What's the difference between them?

STORMY WEATHER

It's a question of geography. Typhoons, hurricanes, and cyclones are all tropical storms that form over the ocean; the difference in the terminology comes from where they happen. A typhoon is a tropical storm in the Northwest Pacific; a hurricane is the same kind of storm but in the Northeast Pacific or the North Atlantic. And if you're in the South Pacific or the Indian Ocean, that storm is a cyclone. To qualify, all of the above must have a wind speed of at least 74 mph.

THE BUTTERFLY EFFECT

These storms depend on the water for their energy and weaken quickly once they blow over land, although frequently not without doing a lot of damage first. What's more, despite the best efforts of meteorologists and ever more sophisticated forecasting models, they remain fairly unpredictable in a longer-term forecast: the precise point at which they arrive on a coast (and thus the point at which they're most powerful) often can't be accurately judged until they're very close. Why? Because a hurricane—or cyclone, or typhoon—has an almost infinite number of variables. When Edward Lorenz came up with chaos theory in the 1960s to explain innate unpredictability, it quickly became known as the butterfly effect, because it included the idea that the flap of a butterfly's wing in Brazil might ultimately give rise to a tornado in Texas. The "butterfly" origins of most U.S. hurricanes, for example, lie far away in the Sahara.

What's worse than one hurricane? Two hurricanes—especially when they run slap-bang into each other. Named the Fujiwhara effect in 1921, after a Japanese meteorologist who was the first to record it, usually the two storms circle one another before the weaker is absorbed into the stronger. This may create a superstorm in the short term, but there's no evidence that combined storms last any longer than their single cousins.

Since 1973, the power of a hurricane has been measured on the Saffir–Simpson scale, which rates it between 1 and 5. At the milder end, a category 1 hurricane has a speed range of 74–95 mph; by the time you get to category 5, the wind speed measures a terrifying 157 mph or more. Experts point out, though, that the damage a hurricane does depends on where they blow— provided they stay out at sea without coming ashore, many hurricanes don't do much harm. Over shallow water, though, they can do a lot of damage to fragile ecosystems such as coral reefs.

THE VERY TINY HURRICANE

The smallest hurricane on record is Storm Marco from 2008, which had winds radiating out only 12 miles from its center, and touched down near Veracruz, Mexico, causing slight flooding. Hurricanes can easily be hundreds of miles long, so Marco qualifies as petite.

21 WHERE DOES WEATHER BEGIN?

Does weather start small (a local shower of rain) and scale up, or start large (a hurricane), and scale down? Or is it a mix of the two?

Put at its absolute simplest, everything to do with the weather starts with the Sun, which warms the air. As the warmer air rises, cooler air moves in below it—and the different air movements cause wind. Different parts of the Earth are warmed to different degrees because their distance from the Sun varies, and varying temperatures affect the water vapor in the atmosphere, ultimately causing clouds and rain. Most of what we experience as weather happens in the troposphere, the level of the atmosphere closest to the Earth (which is why weather forecasters refer to atmospheric conditions), but its causes lie way out in space.

BEYOND THE EARTH

Weather happens a long way away from Earth, too, and can have repercussions light-years from its source. The Sun has an eleven-year weather cycle, for example— we're currently in cycle twenty-five. Solar eruptions, a regular part of this cycle, shoot vast clouds of plasma and vapor out into space, and the resulting geomagnetic storms can have far-reaching effects. From the Earth's point of view, you could see the more dramatic events of the Sun's weather as a series of near-misses. The Carrington event in September 1859 was the last major recorded blip in Sun–Earth relations; outsize solar flares caused a geomagnetic storm so immense that the night lit up like daytime and all the telegraph systems were knocked out. (Telegraph operators, though, were amazed to find that although the storm had blown out telegraph systems, there was so much electricity in the air that they could send communications without them.) The most recent narrow miss was in 2012, when a huge geomagnetic storm just missed Earth—fortunately, because our massively increased reliance on the electrical grid and GPS systems would mean more widespread blowouts, and the possibility of whole cities losing power at a stroke.

WHEN DOES WEATHER BECOME CLIMATE?

If you take weather and add time, you get climate. As the climate scientist John Kennedy tweeted, "Practically speaking: weather's how you choose an outfit, climate's how you choose your wardrobe." Nevertheless, there's long been a gap between the short term of a weather forecast (which meteorologists can make reasonably accurately for up to two weeks ahead) and the long-term views of climate scientists. And there are growing numbers of studies that are working on ways to forecast weather—accurately—a month or two ahead.

22 WHY IS EARTH'S SOIL LAYER SO THIN?

If you've ever tried to grow anything, anywhere, you'll know how important soil quality is. Almost all plants require a little soil to grow, and many need a thick layer of good, fertile soil to be productive. But with soil so essential to life, why is Earth so thin-skinned?

SLOW DEVELOPER

Making soil is an extraordinarily laborious process. And there aren't any shortcuts. It's made from rock fragments mixed in with air, water, and organic matter from plants. To arrive at the end result, slabs of rock must be broken down, over centuries, by weather, occasionally given a geological helping hand by a nearby glacier. Once a rock surface has cracks, plant life takes advantage of them, and roots help to break the stone down further (no surprise to anyone who has ever seen a dandelion growing through a concrete slab). As plants die they add extra organic matter to the mix, and very, very slowly, smaller rock pieces break down further and integrate with organic matter to make soil. In a temperate climate, it can take a thousand years for a single inch of soil to form; it's a little faster in damp, tropical surroundings, but you're still looking at a 500-year wait. Once formed, soil remains "live" and in a constant state of change and development.

Given its thousand-year-per-inch gestation process, it's not surprising that Earth's soil layer—its "skin"—isn't very thick, rarely going deeper than 24 inches.

SOIL MIGHT MAKE YOU HAPPIER

Gardeners have always known that getting your hands in the earth can make you feel better. But it may not just be the effect of physical activity out in the fresh air that is doing you good—new research is increasingly arguing that soil bacteria play a part in making us feel better.

Studies into the human microbiome, the community of bacteria in your gut, have found that it has effects on far more aspects of your health than simply your digestion. And, like humans, soil has its own microbiome. Over the last two decades, a lot of research has been done into whether the microbiome of soil could be used to help human health.

Just one example: In 2004, a group of patients in the late stages of cancer were injected with *Mycobacterium vaccae*, a bacterium commonly found in soil. The experiment was a part of a research study; the hope was that the bacteria might help to boost their immune systems, allowing their bodies to stage a fight against the cancer cells. Sadly, it wasn't successful, but for those injected there was a happier side effect: the patients' moods lifted, their anxiety levels fell, and they found they could think more clearly. Subsequent laboratory studies reinforced the findings and established that (in mice at any rate), *M. vaccae* affected those neurons in the brain that prompted the release of serotonin. In humans, low rates of serotonin can result in depression; it may be, in the future, that we turn to microbiomes of the soil to help with all kinds of health problems.

23 WHY ARE SOME MINERALS SO RARE?

Have you ever heard of cobaltominite? How about fingerite? Or the tongue-twisting metasideronatrite? Probably not; they're among the world's rarest minerals, many of which exist in only the tiniest quantities, found in some of the most far-flung places on Earth.

CLUES FROM THE PAST

Every different mineral is made up of a unique combination of elements, created by varying atmospheric and biological factors. The rarest are rare because the circumstances that formed them were either unique or highly unusual. If it's studied by someone who knows how to read it, every one contains some information about Earth's past. Many minerals don't give up their secrets without a fight, though: some, like Dracula, disappear when exposed to light, while others can simply be washed away in a shower of rain.

At the birth of our own solar system, it's believed that there were only around a dozen minerals; the thousands since added to the list have largely come about as a result of the development of life on Earth, with its myriad interactions between different substances and varying climate conditions. Over 100 new ones are still added to the catalog every year.

THE MINERAL HUNTERS

Of around 5,000 known minerals, only 100 are "common"—that is, they make up a significant, measurable amount of the Earth's crust. All of the rest count as rare, but there are over 2,000 identified minerals that are found in no more than five places anywhere on Earth.

Two scientists, Robert Hazen of the Carnegie Institute in Washington, D.C., and Jesse Ausubel of Rockefeller University in New York, set out to record these rarities formally, producing a catalog in 2016.

At the time, they were asked why they would make a study of such oddities: surely a mineral so rare that the world's supply of it would fit on a fingertip couldn't have much significance? On the contrary, they countered—it's exactly those minute traces from the past that reveal the history of the planet. We share the commonest minerals with many other planets; it's the rarities that reveal the unique qualities of Earth.

"A PERFECT STORM OF RARITY"

Fingerite is perhaps the poster child of rare minerals. In Dr. Hazen's words, "(It's) like a perfect storm of rarity . . . It occurs only on the flanks of the Izalco Volcano in El Salvador—an incredibly dangerous place with superhot fumeroles. It's made of rare elements—vanadium and copper have to exist together, and it forms under an extremely narrow range of conditions . . . And every time it rains, fingerite washes away."

Dr. Hazen has a mineral named after him, and it too is gratifyingly obscure: made up of minute transparent crystals excreted by microbes, hazenite has been found only in Mono Lake in California, forming when the water's phosphorus levels rise.

24 WHY DOES GLOBAL WARMING SOMETIMES MAKE THINGS COLDER?

If the world is growing warmer, why do we still see spells of extremely cold weather, sometimes in areas that aren't used to them?

THE POLAR VORTEX

Global warming causes upset to many systems we've long regarded as established. Cold weather around both the North and South Poles is held in place by a polar vortex, a circuit of cold winds high above the Earth's surface. In recent years, there have been times when the polar vortex of the North Pole has slipped, and when it does, cold air makes its way farther south in the Northern Hemisphere than would normally be the case. In recent winters, the cold-air seeps have caused unprecedented cold snaps in locations as far-ranging as southern Europe, eastern Asia, and the northeastern United States. Why the vortex slips is a question that may have several answers—one theory is that higher temperatures in the stratosphere, high above it, cause it to wobble off course. The Arctic Circle is also warming at twice the rate of the rest of the world; another possible contributing factor may be that there's now less ice to reflect the Sun's energy, and the seas have warmed overall as a result.

THE COLD BLOB

It may sound like a Marvel character, but the Cold Blob is actually an area in the North Atlantic where glacier melt from Greenland is causing a "blob" of icy water. Weather scientists hold it responsible for slowing down the northern part of the Gulf Stream, the warm, fast ocean current that runs up the east coast of the Americas before crossing to Europe.

FIVE QUICK FACTS

1 IT CAN (OCCASIONALLY) RAIN FISH

When a tornado passes over the ocean, its powerful suction sometimes draws fish along with water into its maelstrom and carries them with it over land—then lets them go as the storm's strength starts to wane.

2 SOIL IS HEAVILY POPULATED

Just a single teaspoon of fertile soil holds an astounding number of microorganisms—an estimated billion cells from up to 10,000 different species.

3 WHEN YOUR BREATH FREEZES, IT "WHISPERS"

When temperatures outside go lower than around -60°F, your breath turns to ice crystals as it leaves your mouth. The Yakut people of eastern Siberia call the soft rustling noise that your breath makes as it freezes "the whisper of the stars."

4 MOUNT WASHINGTON BOASTS OF BEING "HOME OF THE WORLD'S WORST WEATHER"

Standing as it does at the meeting of three separate storm paths—from the Gulf, the Atlantic, and the Pacific Northwest—it endures hurricane-force winds for at least 110 days per year.

5 TEENY TARDIGRADES ARE TOUGH

Could they out-survive cockroaches— normally the chart-topping species in lists of top survivalists? These cuddly looking microanimals (just 0.02 inches long) can withstand deep-sea water pressure and high levels of radiation, but they're not good with prolonged heat. Sadly, the roaches still win.

25 COULD THE WHOLE WORLD FLOOD?

Global warming has made extreme flooding events more common, and plenty of ancient legends describe catastrophic floods. Could there ever be a flood in which every part of the world was covered in water?

WATER VS. LAND

If you haven't looked at a globe recently, it may come as a surprise to remember that over 70 percent of the Earth is covered in water. The vast majority of the water on the planet—over 95 percent—is in the oceans. What's left over, frozen at the poles, as groundwater in aquifers, hanging in the atmosphere as water vapor, or free flowing in lakes and rivers, makes up quite an impressive total, but it's still not nearly enough for a full global flood. According to the U.S. Geological Survey, if you added all of it together (including all the melted glaciers and ice caps, and all the water in the atmosphere condensed into rain and falling in one go), it would still represent only around a quarter of the water needed to cover all the land areas on Earth.

WAS NOAH'S FLOOD REAL?

Noah and his ark play their part in probably the most familiar of the flood legends, but could any of the story be true? In the form in which it's described, with the entire world flooded, probably not. But large bodies of water have often moved and changed in geological history. In 2000, two marine scientists looking for evidence of a catastrophic flood in biblical times found complex sediment layers under the present-day Black Sea that seemed to show that over 7,000 years ago its northern part had been fertile farming land around a much smaller body of freshwater. It was sited substantially below the level of the Mediterranean, and separated from it by the Bosphorus, then a thin barrier of solid land. If a landmark weather event raised the level of the Mediterranean, it would have poured over into the farmlands like a vast dam bursting, causing apocalyptic floods for the farming communities below. And although the legend may claim that floods covered the whole world, it has to be remembered that in 7000 BCE known worlds were more local, and therefore smaller.

TWO BY TWO

There are other questions raised by the story of Noah, one of which is how on earth the ark could have floated with all those animals on board. A study in 2014 at the University of Leicester worked out that hypothetically, even loaded with 35,000 pairs of animals (and calculated by load weight only), a boat of the dimensions recorded in Genesis would have been feasibly buoyant. It would have been noisy, and it wouldn't have smelled great, but it would have stayed afloat. How was the equally hypothetical number of animals arrived at? It was argued that the biblical text mentions "kinds" of animals, which was loosely interpreted as a pair of animals from major family groups only. If every species had been included (currently estimated at around 8.7 million, including those as yet undiscovered), the ark would certainly have sunk.

26 WHO WAS THE FIRST ECOLOGIST?

The word *ecology* (*oekologie* in German) was used for the first time in 1869 by Ernst Haeckel, a German doctor, biologist, zoologist, and naturalist. Born in 1834, he was a devoted student of Darwin and the term had its roots in Darwin's theory of evolution.

ART AND NATURE

Haeckel is still best known for the outstanding beauty of his drawings and engravings, published in several volumes, as *Kunstformen der Natur* (*Art Forms in Nature*). Most of his work showed marine life—jellyfish, sea slugs, anemones, and many more creatures—all in exquisite detail. They were combined into volumes of vividly colored and outsized plates, which are still in print and will be recognizable to many people who may never have heard Haeckel's name. Some of his other beliefs—he was an enthusiast for eugenics—have cast an indelible shadow on his scientific legacy, but the idea of *oekologie* took root, and it was soon accepted as a distinct science in its own right.

FROM ECOLOGY TO ECOSYSTEM

The idea of a science that studied the ways in which all organisms work in the natural world steadily developed from Haeckel and Darwin onward. In 1875, the Austrian geologist Eduard Suess was the first to use the word *biosphäre*—anglicized to "biosphere"—to describe the area of Earth, its surface and lower atmosphere, which supports all forms of life. Finally, in 1935, the ecologist Arthur Tansley coined the term *ecosystem*—condensing "ecological" and "system"—to describe both the organisms living in an environment and the environment itself.

THE MISSING LINK

Haeckel didn't actually invent the idea of a "missing link"—an exact point between one stage of evolution and another—but his application of it still persists. It developed in the late eighteenth century as part of the notion of a "Great Chain of Being" in which every life form had its place and each stage of evolution followed neat, progressive connections between species. Enthusiasts for the concept believed that species evolved along a rational path, in which fish evolved into frogs, frogs evolved into lizards, and so on all the way "up" the chain, with humans at the top. Haeckel used it to create a 24-stage process in human evolution, pinpointing stage 23 as the missing link, a precise transitional point between apes and humans. He even gave the "link" a name, *Pithecanthropus alalus* (literally, "the ape-man without language"), although there was no proof that it had ever existed. Despite the fact that the missing link has long been dismissed by scholars, the term has persisted as a sort of nonscientific shorthand, and a number of fossil contenders have been lined up, from "Java man" found in Indonesia in 1891, to a series of skeletons found in South Africa between 2008 and 2010, named *Australopithecus sediba*.

27 COULD A BUTTERFLY REALLY CAUSE A TORNADO?

There is a saying that a butterfly flapping its wings in Brazil could, in time, cause a tornado in Texas. But is this really true? Well, yes and no. But mostly no.

CHAOS THEORY

Edward Lorenz was a pioneer of chaos theory. He was working on statistical modeling in weather forecasting and he struggled to make the numbers work. He found that by making tiny, seemingly inconsequential changes to the initial conditions in his models, the outcomes would vary enormously. His exploration of this idea eventually became known as the butterfly effect.

We often think of the world as simple cause and effect, and that we can predict accurately, often with the help of computers, exactly what's going on. Chaos theory adds more complexity; it is present in any system where small changes make a big difference. The weather is only one such system, though it explains why the weather reporter is so often wrong.

Chaotic behavior is very common in the physical world and touches aspects of physics, chemistry, biology, and mathematics. It is possible that, in time, with increasing computing power, chaos theory will simply be very difficult math— but for now at least it means that so many things in our universe are essentially unpredictable.

THE BUTTERFLY'S TORNADO

The butterfly effect theory proposes that it is possible that the changes in air pressure caused by a butterfly's wings flapping could be the start of a long and complicated chain of reactions that eventually forms a tornado. However, the chain would be so long and complex, with so many other influencing factors, it could hardly be attributed to just the flapping of some wings.

QUIZ
EARTH MATTERS

The planet we stand on has many unsolved mysteries. But we do have some answers—try out your knowledge here in this chapter's quiz.

QUESTIONS:

1. Where do typhoons form?

2. What is the Saffir–Simpson scale?

3. How long does the Sun's weather cycle last?

4. Where is the mineral fingerite found?

5. What is the Cold Blob?

6. How much of the Earth is currently covered in water?

7. What is the Fujiwhara effect?

8. How long does an inch of soil take to make?

9. How many new minerals are discovered annually?

10. What is the smallest hurricane currently on record?

Turn to page 245 for the answers.

FORCES AND MATERIALS

WHY CAN'T WE GO FASTER THAN THE SPEED OF LIGHT?

There is a universal speed limit. Nothing can go faster than the speed of light (often written as "c"), which is 983,571,056 feet per second. It doesn't matter how many rockets you strap to something, you can't get any faster. This is because as something gets faster, it also gets heavier.

IT'S ALL DOWN TO INERTIA

This comes from Isaac Newton's principle of inertia, which, simply put, states that if you want to accelerate something, you need to use energy to do it. This is intuitive: If you give a wooden block a push across a table then you are using energy to make the block move. Then if you do the same but with increasingly heavy objects, they get harder and harder to move and you need to use more energy, until you get to something so heavy you can't push it. So the heavier something is, the more energy it takes to speed it up. We are limited by how much energy we can produce.

SCIENCE FICTION EFFECTS

Traveling at super high speeds does some mind-bending things. Not only does it make you heavier, but if you were traveling very fast you would get shorter and your perception of time would slow down relative to the universe around you. This leads to all sorts of strange effects that you might think only matter in science fiction—but some of them are very real and need to be taken into consideration for things like GPS satellites and detection of deep space objects.

Muons are short-lived particles that are made in our atmosphere by cosmic rays. They travel at near light speeds, but even going that fast they decay so quickly they should never reach the ground—the only reason they do is that because they travel so fast, the time they experience slows down, meaning they last long enough to reach ground-based detectors.

INCREASING WEIGHT WITH SPEED

The full answer to why you get heavier when you get faster is complicated and involves some difficult math, but it can be simplified to the following equation:

$$m = m_0 \times \frac{1}{\sqrt{1 - \frac{v^2}{c^2}}}$$

What this equation tells us is that the mass of an object (m) is based on its "normal" mass (m_0), our speed (v), and the speed of light (c). At normal speeds that we experience in day-to-day life, v is so small compared to c that m and m_0 are basically the same. It's only when you get to about 25 percent the speed of light (still a whopping 245,892,762 feet per second) that the effects become even slightly noticeable. From this point, as the speed increases, v^2 gets closer to c^2, which makes v^2/c^2 get closer to 1. This makes the number that m_0 is multiplied by bigger and bigger, until you reach a point where v=c, and this makes the number you multiply m_0 by infinity, meaning that m, your mass, will be infinite!

So if you want to accelerate an object past the speed of light, you will need an infinite amount of energy to speed up your infinitely heavy object. As this is not possible, it means we can't go faster than the speed of light.

30 WHAT MAKES NUCLEAR WASTE UNSAFE FOR MILLENNIA?

Nuclear fission can produce huge amounts of energy without using much fuel, so it might seem like the solution to the energy crisis. The problem, however, is that the waste created is dangerously radioactive and will remain so for thousands of years because of its radioactive half-life.

HALF-LIFE

If you were to take a lump of radioactive material with, say, 2,000 radioactive uranium atoms in it, over time they would spontaneously emit some form of radioactivity and decay, becoming inert. The reason this happens is due to a very complicated quantum effect, but it leaves us with an interesting pattern. Over a set period of time, known as a half-life, roughly half of the atoms in a material will decay. In our uranium example, there will be only 1,000 radioactive atoms left. When the same time period passes again, half of the remaining atoms will decay, leaving only 500. Then again after another half-life there will

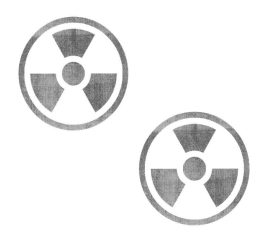

be just 250, and so on and so on. It can be difficult to wrap your head around why it happens, but it's like flipping a coin for every atom each half-life. There's a 50/50 chance that any individual atom will decay each time, so roughly half of them do. This half-life effect means that while the radioactivity of nuclear material will weaken over time, it will take a long time for it to decay to the point where it is safe for humans to be around.

NUCLEAR WASTE

Different nuclear materials decay at a different rate. Some that are used for medical purposes have a half-life of only a few minutes, but those that come out of nuclear power plants tend to last much longer. Uranium fission produces elements such as cesium-137 and strontium-90, which have half-lives of about 30 years, and plutonium fission can produce plutonium-239, which has a half-life of 24,000 years!

RAY CATS

One of the problems with nuclear waste is keeping it safe—not just now but in the future. Much of it is stored in vast underground vaults that are sealed off, but in 10,000 years any knowledge of these places may be lost, all written and spoken languages may be dead, and even the idea of what radiation is may be gone. So how can we communicate to future generations that these are dangerous places to be avoided? One of the most interesting ideas proposed is to genetically engineer "ray cats," which would be the same as normal cats except that they would change color or glow where there is radiation present, and to then create a legend that if the cats change color, you should leave immediately. That way, even if the words we use change, as long as the knowledge is passed down that if the cats glow, you should stay away, the future people of Earth could stay safe.

31 WHY DO THINGS MELT AT DIFFERENT TEMPERATURES?

You can melt just about anything. But the temperature at which stuff melts can differ wildly. This is because melting happens when a material has been heated up enough for the atomic bonds that keep it solid to be broken. Materials have different types and strengths of bonds, so different temperatures are needed to break them.

STRONGER AND WEAKER BONDS

There are many different types of bonds that can occur. Some rely on atoms or molecules being polar, which makes them act almost like tiny magnets, which can then attract each other. Others form when atoms are able to share electrons between themselves, or where the act of losing or gaining the electrons causes them to become oppositely charged (like opposing ends of a magnet), which then causes bonds to be created. Of course, as well as the type of bonds, which atoms are forming them makes a big difference. Larger, denser atoms often make weaker bonds but may also be able to support more than one type of bond. Atoms bond to form molecules, and these molecules can also bond to other molecules, creating more bonds that will need to be broken in order to melt the material.

EXTREME BONDS

The highest known melting point belongs to tantalum hafnium carbide (Ta_4HfC_5), which melts at a scorching 6,192°F, whereas the lowest melting point is helium at -458°F, which is only one degree or so above absolute zero.

FIVE QUICK FACTS

1

THE HARDEST SUBSTANCE ISN'T DIAMOND

In the last few years, diamond has been overtaken as the Earth's hardest substance by a new superhard carbon crystal—named "nanocrystalline diamond balls"—made by binding tiny crystals with a layer of graphene, just one atom thick.

2

ALTHOUGH WE SAY "HEAVY AS LEAD," IT'S FAR FROM THE WEIGHTIEST METAL

That honor goes to osmium, which is the densest element on the periodic table, and is more than twice as heavy.

3

BRIDGMANITE, EARTH'S MOST ABUNDANT MINERAL, WAS ONLY NAMED IN 2014

Although most of it is located under the planet's crust, it makes up an estimated 38 percent of Earth's overall volume.

4

KEEN STARGAZERS SHOULD HEAD TO THE CANARY ISLANDS

In a research study into global light pollution carried out in 2021, the Roque de los Muchachos Observatory on the Spanish island of La Palma won out over other dark spots such the Sierra Nevada in the United States, and the Extremadura in Spain.

5

THE GREAT UNCONFORMITY REVEALS THAT A HUGE CHUNK OF TIME IS MISSING

Generally layers have been added to the Earth's surface relatively consistently. But there's an unexplained jump between layers from the Cambrian era, 540 million years ago, and the layer beneath it, which is a billion years old. The so-called Great Unconformity is a geologist's escape room puzzle.

32 WHAT CAN YOU USE TO FLOAT A TRAIN?

You might think that a floating train is some crazy science fiction idea, but it's already happening here on Earth today! Various floating trains have been built in Germany, South Korea, and Japan since the 1960s. These trains are able to float using the repelling power of magnets.

OLDER TRAINS

There have been lots of maglev (magnetic levitation) trains, but they have often been short-lived, used mostly for expos and only traveling short distances. The only commercially running one today is the Incheon Airport Maglev in South Korea, which shuttles to and from the airport. The Incheon Airport Maglev and all maglev trains invented before it float using electromagnetic suspension. This is where the bottom of the train is wrapped around a metal track. The train contains carefully controlled electromagnets that allow it to produce a magnetic field with the same charge as the track, keeping it floating just above the metal track.

WHY FLOAT A TRAIN?

Floating trains come with a number of benefits. One of the most obvious is that because the train isn't touching the track, it won't be subject to friction. This means that a floating train is able to move much faster than conventional ones. They can also be made lighter than normal trains and don't put as much strain on the tracks, bringing track maintenance costs down.

It is also very easy to make a maglev train completely electric. Maglev trains don't have wheels to turn; instead, they use their magnetic abilities for forward propulsion, and this is much easier to do with electricity, rather than a gasoline or diesel engine. This in turn is better for the environment and again keeps the costs down.

FLOATING TRAINS OF THE FUTURE

Future floating trains such as the Japanese Chūō Shinkansen will instead use the properties of superconductors. Superconductors are very special materials that, when cooled down to very low temperatures of at least -328°F, will cease to have any electrical resistance at all. As a superconductor is cooled down into its superconducting state, it also takes on another property known as the Meissner effect, which causes the pinning of magnetic fields. What this means is that if a magnet is held above the superconductor as it cools, then when the process is complete it will float there for as long as the superconductor is kept cold.

OTHER USES OF SUPERCONDUCTORS

While the role of a superconductor in floating-train technology is very exciting and useful, its zero-resistance quality is much more interesting. It will allow for better electronics that are faster and don't heat up, which is a major technological problem at the moment. They could allow for better power transfer with less loss, reducing the amount of power that is needed to be produced. They also have uses in medicine, such as in MRIs, and in various scientific experiments, such as in the Large Hadron Collider at CERN. The major problem with superconductors is that all the known ones still need to be kept very, very cold, so the race is on to find a "room temperature superconductor"; if one is ever found, it will revolutionize our world.

33 WHY DO DOOR HANDLES KEEP GIVING ME ELECTRIC SHOCKS?

It's a common problem. You reach for a door handle to open the door and suddenly you get a sharp, painful electric shock. Maybe it's a stair rail or a comb instead, but the point is you keep getting shocked. These things can give you electric shocks because of a buildup of static.

THE STATIC EFFECT

Electricity is simply a flow of electrons from one place to another. In circuits a battery is used to push the electrons around a system, but electricity can happen naturally in the form of electrical shocks. Shocks start with static—the buildup of an electrical charge in a place. This can happen when there are too many or too few electrons. The buildup of electrons causes the object to become charged with electrical energy known as static. This static energy can then be discharged when it comes into contact with something else. It may either absorb lots of electrons or release them very quickly, and when it does so this causes an electric shock.

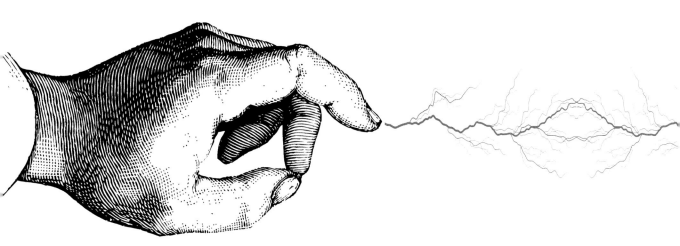

SIMPLY SHOCKING

You are constantly having tiny unnoticeable shocks as electrons move between you and the environment, so why can you sometimes really feel them? To reach a point where you can feel it, you have to build up a lot of static. One of the most common ways static is made is when things rub together, because electrons can be knocked off one of the materials.

We can generate static in our bodies when we move around in our clothes or scuff our feet on carpets. Eventually, when we touch something made of metal—which is able to move electrons around very quickly—this can result in a painful electric shock. We often get shocks from door handles because the rubber soles of shoes keep electrons from moving between us and the ground, so it's only when we touch the metal with our hand that the shock can take place. If you find yourself getting lots of shocks, you might want to avoid wearing clothing made of materials (like wool) that produce a lot of static.

HAIR RAISING

You may have put your hands on a Van de Graaff generator before. When you do this, your hair starts to stand up on end. This is because the bell becomes positively charged as an internal belt rubs the metal and strips away the electrons. When you put your hands on it, your electrons flow to take their place and your body also becomes positively charged. This means that each of your strands of hair becomes charged with static electricity. All of this static is positively charged, and so each hair pushes away all the other hairs until they're all pushed out and floating—just like when you try to push two alike poles of a magnet together.

34 CAN I MAKE A DIAMOND?

You may have been led to believe that diamonds are rare and precious. Some sort of magical stone forged only in the heart of the world. But diamonds are just specially arranged carbon atoms, and it is possible to make one in a lab.

REARRANGING CARBON

In the element carbon, each atom allows four chemical bonds to it. Because there are multiple bonds available, there are lots of different ways that the atoms can be put together. Humans are largely composed of carbon, but our carbon molecules look very different from the largely unorganized arrangement within the carbon molecules in coal. Carbon can also be structured to make graphite for pencils, and single layers of graphite are a material in their own right, graphene.

Diamonds are just another arrangement of carbon. They are made of carbon atoms arranged in a tetrahedron shape. These tetrahedrons then build onto each other, creating the very hard crystal structure we know as diamonds.

MAKING DIAMONDS

Diamonds are made naturally in the Earth's crust by the large pressure and high temperatures the carbon is exposed to. It is possible to make synthetic diamonds by recreating similar conditions with industrial machinery, applying huge pressure to press the atoms into shape and subjecting them to scorching temperatures to burn away all but the toughest diamond bonds. Synthetic diamonds are functionally identical to the ones found in the ground, but they are usually made to be much smaller to keep costs down. These diamonds are often used in construction tools for their hardness, and in some electronic devices. Some even make their way into jewelry.

35 WHY DO THINGS EXPAND WHEN THEY GET WARM?

We know intuitively that when things get hot they swell in size. Doors might stick, pavements crack, and bridges have to be built to allow for it. But why this happens isn't immediately obvious. The reason things get bigger with heat is that they have more energy.

HOW DOES MORE ENERGY MAKE IT BIGGER?

Temperature is really a measurement of how much the atoms in something are moving around. The hotter something is, the faster the atoms are vibrating or moving around. So as you heat something up, each atom will move more. The increased movement of atoms means that it takes up more space. While each individual atom might only take up a tiny amount more space, when this happens to all of them in an object like a rock, it will cause a noticeable amount of expansion.

FREEZING EXPANSION

A select few materials also expand when they freeze. You may have discovered this yourself by putting a full bottle of water in the freezer and opening it up later to discover the bottle has cracked. In liquid form, all the atoms in a substance are able to slosh around freely, but when some chemicals turn into a solid, they settle into crystalline structures. These structures push individual atoms farther apart from each other and lock them in place, making the material as a whole expand.

36 WHY CAN'T WE UNTOAST BREAD?

This is perhaps one of the most important and fundamental questions we can possibly ask. In answering it, we explore the very nature of the universe itself: where it came from, where it's going, and how exactly it will get there. This question (it may not surprise you) isn't really about toast. But we can't untoast bread because of entropy.

ONCE TOAST, ALWAYS TOAST

When you toast bread, a number of things occur. There is a series of reactions as the amino acids and sugars are reacted together, causing them to mix and brown. The sugars in the bread can also caramelize, which makes them melt and spread out. All of this increases the entropy in the toast and, because entropy only ever goes forward, you can't undo it.

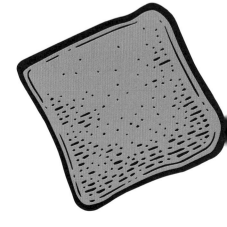

THE ARROW OF TIME

Entropy is defined as "the measure of a system's thermal energy per unit temperature that is unavailable for doing useful work." To the lay reader, however, this might not be particularly useful in understanding what it actually is. It is often called a measure of disorder and it has one key property: Entropy is always increasing. Think of a block of ice dropped into a pot of hot water. At the start, the ice atoms are relatively ordered—cold in one part, hot in another. But over time the temperatures will become the same and the atoms originally in the ice will melt and mix with the rest in the pot. Over time, the disorder in the system has increased.

ORDER VS. COMPLEXITY

If entropy is only increasing, and things are getting less ordered and more chaotic, then how can organized things like stars, planets, and even human beings form? Just because entropy overall must always

CHAOS AND THE UNIVERSE

Entropy is sometimes called (somewhat grandiosely) the "Arrow of Time," because it is the only thing that can only ever go in one direction, always getting bigger. At the beginning of the universe, everything was in a near-perfect order. Since then it's been getting more chaotic, the amount of disorder has been ever increasing, and every process that happens in the universe serves to increase the amount of entropy.

increase, that doesn't mean it isn't able to decrease in some places. It's like how making ice in a freezer will decrease the entropy in the water but the freezer will proportionally increase entropy outside the device. Complexity, however, is very different from order. In our previous example, as the ice and water begin to mix, it becomes increasingly complex, as to define the system properly would be increasingly difficult, until eventually it all becomes just water again—so complexity arises as part of the overall increase in entropy. In a sense, our existence is probably an effect of entropy.

37 WHY DOES METAL CONDUCT ELECTRICITY, BUT WOOD DOESN'T?

If you've ever played around with electrical components, you will have noticed that some materials are able to conduct electricity, but others, such as wood, can't. The difference comes down to the material's electrical resistance and the amount of free electrons available.

ELECTRICAL RESISTANCE

Electricity is a flow of electrons through some kind of system, like a wire, but electricity is always hampered by resistance. Imagine electrons passing through an object as a large group of runners traveling through a dense forest. Different materials will have higher or lower resistances, which is like the trees being more abundant and closer together, meaning that the runners have to slow down or may even crash, making it harder for them to pass through quickly.

FREE ELECTRONS

Metals undergo a special type of bonding known as metallic bonding (it's what makes a metal a metal). One of the results of this is that all the outer electrons around each atom essentially become bonded to the material as a whole, rather than to any individual atom. This means that when you pass a voltage through a metal, the outer electrons from the metal itself are able to flow, making it easier to carry a current.

DIFFERENT METALS, DIFFERENT FLOW

Not all metals conduct equally. Because of the variation in their internal structure, the resistance can be increased and different metallic elements have different numbers of outer electrons. All metals can conduct electricity but some, such as gold and copper, are better than others, such as aluminum and titanium.

QUIZ
FORCES AND MATERIALS

Try not to force the issue, but can you remember any of the material from this chapter?

QUESTIONS:

1. Which element has the lowest melting point?

2. What does the term "half-life" mean, atomically speaking?

3. What would be different about a ray cat?

4. What causes static?

5. What physical effects might travel at superhigh speeds have on you?

6. How can you make magnets multiply?

7. What does Newton's principle of inertia state?

8. Why aren't lab-made diamonds larger?

9. What happens to water when it freezes?

10. Where would you find the world's first commercial floating train?

Turn to page 245 for the answers.

38 WHICH DINOSAUR HAD THE MOST TEETH?

If you grew up watching movies that starred mostly meat-eating dinosaurs, with plenty of fearsome teeth shown in terrifying close-up, dripping with gore, you might be surprised to hear that the species with the largest number of teeth rarely tackled anything more challenging than the tough stems of early plants.

HOLY COW!

If it came to a tooth count, the prehistoric plant eaters are clear winners. There's Nigersaurus, for example, a 45-foot-plus member of the long-necked sauropod family from the early Cretaceous period, who browsed ancient vegetation between 121 and 99 million years ago. Nicknamed the "Mesozoic cow" by Paul Sereno, the paleontologist who discovered some of the most extensive of its fossilized remains, Nigersaurus had around 500 needle-shaped teeth, but only used some of them at a time. Its reserve sets were stacked vertically in its broad-fronted jaw, ready to be pushed into service as soon as the teeth directly above or below fell out.

CHEWING MACHINES

Most of us (with a mere thirty-two teeth each) might be impressed by Nigersaurus's total, but even that count is totally overshadowed by the dental arrangements of the hadrosaurs, a younger (Late Cretaceous) group of dinosaurs that lived around 65 million years ago. Although hadrosaurs were dubbed "duck-billed" by early fossil hunters, the "bill" was really more like a strong, grooved beak. Their diet was herbivorous—believed to have been a tough mixture of branches, leaves, and rough bark—and not only did hadrosaurs have up to 1,400 teeth to cope with it, but each tooth changed function in its brief lifetime (each hadrosaur tooth is believed to have lasted only a couple of weeks). Sharp teeth started out at the front of the mouth, probably where they could be used to cut off manageable chunks of plants. As their surfaces wore down, the teeth themselves shifted sideways in the jaw, and their now-blunter surfaces were used to grind the plant pulp to a softer, more digestible mass. Look at the fossil jaw of a hadrosaur and you'll see a wide band of teeth, with sharper surfaces at one edge, graduating to flatter surfaces, perfect for grinding. As the blunter teeth fell out, they were replaced from the next row along.

Hadrosaurs may not have been particularly sophisticated thinkers, but they were seriously good at chewing. When a team of scientists made a detailed study of their teeth and jaws, it was established that hadrosaurs had been more efficient masticators than most of today's large mammals, and the team leader, Gregory Erickson of Florida State University, dubbed them "walking pulp mills."

KILLER SMILES

In contrast to the toothy herbivores, the (in)famously carnivorous *Tyrannosaurus rex* had only sixty teeth in its mouth. They were big, though! *T. rex* holds the record for the largest dinosaur teeth, with some fossilized examples measuring around 12 inches long.

39 HOW LONG DO MOST SPECIES LAST BEFORE THEY BECOME EXTINCT?

Despite the negative ideas attached to the word, extinction is a natural event—99 percent of the 4 billion-odd species that have lived on Earth since it came into being are already extinct. If they weren't, plesiosaurs would still be swimming in the oceans and diplodocuses would be browsing scrubland. But many people believe that we're currently in the middle of the sixth mass extinction event in Earth's recorded history. What's the "natural" rate of extinction, and how is it judged?

AN INFINITE TIMESCALE

People tend to think in time divided into decades or, at a pinch, centuries, so it can be hard to take in just how short the time humans have lived on the Earth is. When it comes to evolution, you have to adjust your thoughts: it's measured in millions of years, a scale that is hard to imagine. Paleobiologists have made an estimate of a "normal" extinction rate as being 10 percent of species over a million-year period, rising to 30 percent over 10 million years and 65 percent every 100 million years.

SURVIVAL OF THE . . . OLDEST?

Occasionally extinction is declared too soon. The West Indian Ocean coelacanth (*Latimeria chalumnae*), a bony fish around 6 feet long, previously known only through fossil specimens, is probably the most famous survivor into the modern age. Thought to have been wiped out in the last mass extinction, a specimen was caught off South Africa in 1938, with a second species identified a full fifty years later. It's highly vulnerable, but it's already lived for more than 65 million years.

A mass extinction, on the other hand, is a period in which three-quarters or more species on Earth become extinct over two million years or less (before you relax, in geological terms two million years is a very short time indeed). There have been five we know of in the past, the earliest around 440 million years ago, the most recent (the one that finished off the dinosaurs) a mere 65 million years ago. Their causes have varied, but in every case, a vast number of species died out because their habitats no longer supported them.

THE HUMAN IMPACT

Are we in the middle of the sixth? It seems likely. The International Union for Conservation of Nature (IUCN) Red List, the global record of species under threat from extinction, has calculated that over the last 500 years around 900 species have been lost—which is a considerably faster rate than in "normal" times. That's the ones we know about, so is likely an underestimate. We know that species that have never been identified are going extinct, sometimes in large numbers, as complex habitats are being destroyed. And although we tend to associate "modern" extinction with today's problems, there's evidence that as far back as 50,000 years ago, the arrival of humans in an area predicted the extinction of its larger animals. The European lion, the American mastodon, the giant bison, and the cave bear are just a few species that didn't survive long after the arrival of people.

There's some reassurance in this picture and it's that Earth will probably carry on just fine. It has coped with many billions of years of change, and it's almost certain that it will survive humanity, too.

40 WHERE'S THE WORLD'S MOST FERTILE SOIL?

Soil needs a number of ingredients to be fertile—plenty of organic matter, good aeration, and a rich ecosystem (around a quarter of known species on Earth live in the soil, including fungi, protozoa, algae, and bacteria). To rate soils, the U.S. Department of Agriculture (USDA) has developed a taxonomy of twelve different types of soil, rating their chemical composition and degrees of fertility.

BLACK GOLD

The soil category that generally rates most highly are the Mollisols, also called black soils or prairie soils, which occupy around 7 percent of Earth's ice-free land surface. They are dark-colored and rich, with very high quantities of organic matter, supporting vast numbers of microorganisms, and in the main they developed under grasslands or steppes, benefiting from the extensive root systems of tough prairie grasses. They're found extensively in the United States and Canada, southern Russia, Ukraine, and northeast China, as well as in the traditional grasslands of South America. Unfortunately, they've proved to be victims of their own success in recent decades, increasingly suffering from over-intensive farming methods.

EARTHWORM INVADERS

Most farmers and gardeners know that wormy soil is a good thing. Not in the Arctic, though, where the advance of the earthworm is considered a serious threat to biodiversity. Traditionally believed to be too cold for worms, the soil there is covered with a thin layer of plant debris, which layers up over the years, creating a unique habitat. Now that the Arctic is warming up, earthworms have arrived and they are pulling leaf litter into the ground and raising the soil's organic content while removing its protective layer. Jonatan Klaminder, an environmental scientist at Umeå University in Sweden, has raised concerns over this threat to very rare and remote environments. In this case, earthworms, he says, are the canaries in the ecological "cold mine."

FIVE QUICK FACTS

1 **T. REX SEEMS TO HAVE BEEN A SCRAPPY TEENAGER**

Studies of the fossils of immature *T. rex* specimens show numerous fresh bite marks with the dental patterns of . . . other tyrannosaurs. The bones of infants don't have bite marks, and the bones of mature adults show old bites, not fresh ones.

2 **A 44-MILLION-YEAR-OLD FOSSIL SHOWS THAT WHALES ONCE WALKED**

Phiomicetus anubis, a four-legged whale ancestor, was discovered in Egypt's Western Desert, and is believed to have been semi-aquatic, living partly on land, partly in the sea.

3 **THE PORTUGUESE MAN O'WAR JELLYFISH ISN'T A SINGLE ENTITY**

It's a colony, called a siphonophore, made up of tiny, dependent separate organisms, hydrozoa, whose ancestors were already living in the ocean over 540 million years ago.

4 **THE FIRST RECORDED FOSSIL FINDING WAS MADE IN 1676**

From contemporary drawings, it's been tentatively identified as the thighbone of Megalosaurus; at the time, Robert Plot, the professor of chemistry to whom the fossil was given, believed that it was the leg bone of an ancient human giant.

5 **LATIN NAMES AREN'T ALWAYS SERIOUS**

The Linnaen binomial system means even the smallest invertebrate may have an imposing name. Eccentricities out there include species named after David Bowie (a huntsman spider, *Heteropoda davidbowie*) and Spongebob Squarepants (a fungus, *Spongiforma squarepantsii*), and the wasp christened *Aha ha*—short but expressive.

41 HOW DOES CARBON DATING WORK?

You've probably seen a museum exhibit that shows a fossil or an ancient tree stump and a small plaque that dates it as millions of years old. Chances are that the given age was ascertained through radiometric dating. Carbon dating is one form of radiometric dating that uses the radioactive decay of carbon to analyze samples.

CARBON DATING

Every living thing contains a little bit of the element carbon-14. We take it in through the air that we breathe and it ends up in our bodies. Living organisms take in and use up carbon-14 at about the same rate, so there's always roughly the same amount in a body; however, when the organism dies, it stops taking in more carbon-14.

Carbon-14 is radioactive. A radioactive substance decays over time; it also decays at a constant rate because of its half-life (see page 72). Scientists are able to use the half-life of carbon-14 to determine how long ago an organism died. They do this by comparing how much carbon-14 you would expect to see in a living version of the

FINDING FAKES

You might think that radiometrics would only be good for finding out fake fossils, but using half-lives to date objects is also used to find fake art! Between 1945 and 1963, there were a lot of nuclear tests across the world. One of the results of this is that many things made today have more radioactive material in them than those made before that time period. One of these things is a binding agent used to make paints. Any paintings made after 1963 have more radioactive material in them, and this is used to help identify fake paintings pretending to be from before that period.

organism against how much there is in the sample you're analyzing. If there's half as much, then it died about 5,715 years ago; if there's a quarter, then it died about 11,430 years ago; an eighth means 17,145 years ago, and so on.

BEYOND CARBON

Carbon dating is quite limited. Because of the relatively short half-life, it can only reliably be used to measure so far into the past. After about 50,000 years, or just under nine half-lives, the amount of carbon-14 becomes very small, making it hard to detect. Fortunately, there are lots of other radioactive materials found in rocks and other substances that can be used to date objects such as fossils. The differing half-lives means that they can be used for lots of different purposes.

The longer a half-life is, the further back the dating method can be used (some, like thorium and samarium, have half-lives longer than the age of the universe), but the less precise it becomes. Carbon dating can get the date right to within about sixty years. Samarium dating, on the other hand, has no upper limit in age but it only gives a date to within the nearest 20,000,000 years. Bioarchaeologists therefore have to use a mixture of different types of radiometric dating, depending on the approximate age of a sample, along with other techniques to try and determine a rough date for the creation of the object.

42 ARE HUMANS REALLY NAKED APES?

Well, it's certainly true that humans are apes—with the much-quoted fact that we share 98.8 percent of our DNA with chimpanzees, it would be hard to make a case against it. But if we're so closely related, why is it that humans, alone among primates, don't have fur?

THREE THEORIES

All kinds of ideas have been put forward, although no single argument has yet won all-around agreement from both archaeologists and paleontologists. There are three main threads of debate.

The first is that we were originally ocean-dwellers. This one argues that, at some time during their evolution, humans were at least partially aquatic. While there are fully ocean-going mammals, such as whales and dolphins, that don't have hair, seals—at home in and out of the water—do; semi-aquatic humans might have grown sleeker, but there's no convincing reason to think they would have lost their hair altogether.

TEMPERATURE CONTROL

Perhaps the strongest argument that's been advanced for humans being the only non-hairy primates is the ability to sweat—and people sweat heavily. Sweating is the most efficient way to cool down, but it doesn't work fast enough to be effective on skin that is thickly covered in hair.

PLANET OF THE APES

If you believe in evolution, you already accept that you and all other humans are apes (or primates, perhaps—it sounds so much more civilized). But it's still hard to take in quite how closely we're related. Nonetheless, many abilities that are widely accepted as exclusively human are actually shared by other primates. Opposable thumbs? Monkeys and apes have them. Laughter? Chimpanzees and some other primates laugh if they're tickled (although, to be fair, it seems that rats do, too). Walking on two legs? There's evidence that some species of apes are comfortable walking bipedally at least some of the time. Tool use and making shelters are also primate skills, not just human ones. We may not be quite as exclusive as we think we are.

Humans seem to have begun to hunt on open grasslands around 3 million years ago, and unshaded, prairie-like conditions may have meant that they needed to shed hair and begin to sweat effectively. If this was the case, becoming "naked apes" would have been a practical evolutionary solution.

PEST ELIMINATION

The third, recent, argument is that humans lost their hair as a way of ridding themselves of pests and sickness. Ticks and lice, and the—often serious—diseases they spread don't settle on hairless skin; it's too hard to get purchase, and too easy for the target to brush them off. Once humans had smooth, naked hides, the theory goes, they were also largely relieved of the problems of many skin parasites.

Which, if any, of the three theories is the right one? While the intriguing aquatic idea has few supporters, each of the others has strong adherents. Although, of course, there's still the very real possibility that some altogether new idea may arrive to challenge or eclipse them.

43 WHY DOES DNA DEGRADE?

It was probably the success of *Jurassic Park* back
in 1993 that pushed a question that up to this point
had preoccupied only serious scientists into the
mainstream: Could you really make a new dinosaur
out of preserved dino DNA?

YOUR DNA IS BROKEN

The idea of dinosaur blood taken from
insects preserved in amber sounds
extravagant, but it was actually based on a
series of experiments and discoveries made
in the same decade as the movie. Although

some scientists claimed that they had
managed to extract DNA that was
120 million years old, gradually these
claims were debunked. Although the DNA
molecule is huge and, in chemical terms at
least, simple, it relies on many links, and

it's these that break as the molecule ages. When enough links have broken, the DNA is technically useless—Helen Pilcher, author of a book about the "new science" of de-extinction, memorably wrote that it would be like "trying to construct the 5,195-piece Lego Star Wars Millennium Falcon from just a few bricks and the picture on the box." As soon as something living dies, its DNA is susceptible to damage. It's affected by enzymes from other organisms it comes into contact with, and oxygen, water, and sunlight can all damage and break the ladder-rung strands that are key to its structure.

OLD, BUT STILL USEFUL

So how long can scientists expect DNA to live? In 2012, an Australian study into DNA taken from the bones of a huge extinct bird, the moa, showed that it had a "half-life" of 521 years (meaning that by the end of that time frame, half of its links would still be complete). In the (relatively) complex math of science, this meant that half of the remaining half would take a further 521 years to degrade, half of that remaining half another 521 years, and so on—resulting in a calculation that determined that it would take 6.8 million years for DNA to be completely destroyed. If this calculation proves correct, it means that it's not an impossibility for feasible DNA to be found in samples around, say, a million years old.

THE DIFFICULTIES OF DNA

Originally, the unique "fingerprint" of an individual's DNA was thought to be the magic bullet that would make criminal convictions foolproof. It's only over time—the first crime to be solved by DNA profiling (the murder of Dawn Ashworth, a British teenager) was in 1986—that things have turned more complicated. As tests for DNA have evolved to be more and more sophisticated, it's become much harder for the presence of small quantities to unequivocally prove something—for example, it's been found that it's possible to transfer DNA from one garment to another in the course of a washing-machine wash, a process that an amateur might assume would destroy the DNA altogether. And now that minute amounts of DNA can be reliably identified, it's also more credible that they could have been deposited innocently: handed over on spare change, for example. Whatever its future, it seems certain that DNA evidence will become more rather than less complex to handle in the future.

44 IS DEATH THE NORM FOR EVERY LIVING THING?

Threescore years and ten may be the biblical span allotted to humans, and many other familiar animals and plants have quite finite lives, but if you cast your net wider, you can find some kinds of life that don't seem to play by what you thought were the rules.

TOUGH AS A TARDIGRADE

Tardigrades, tiny animals that live in a variety of mostly damp habitats, have a cartoonish name and look like Pokémon personalities if you put them under the microscope—which you will have to do if you want a detailed look, as few of the thousand or so different species of tardigrades grow longer than a millimeter. But their cute appearance (solid bodies, squashed-up faces, and eight legs with clawed feet, all of which have led to their colloquial names—water bears or moss piglets—is misleading: tardigrades are one of the toughest life-forms in the world. Although their favored environment is a damp one—many species live in sediment at the bottom of rivers or lakes—tests have found that they seem to be very nearly indestructible in a wide variety of habitats and situations (they've even survived trips in space).

One of their special talents is to self-dehydrate, bringing metabolic processes to a sharp halt to achieve a near-death state called cryptobiosis, which enables them to survive apparently indefinitely until conditions become more favorable.

ACHIEVING IMMORTALITY

Other non-dying oddities include *Turritopsis dohrnii,* the immortal jellyfish, which can grow younger as well as older. For most of us, reproduction is a stop on the path to redundancy and, ultimately, death, but the immortal jellyfish lives up to its name by reproducing, then reverting to a polyp, the immature state in jellyfish. It then grows up all over again, reproduces, and repeats the cycle. Ad nauseam.

THE UNDEAD

Viruses may be a slightly unfair entry on the list of immortals, as there's long been a scientific debate about whether or not they're alive in the first place. Some consider them simply lineups of chemicals—but they challenge that definition when they hijack cells and get active. The cells (called hosts when they've become unwitting homes to a virus) are subjected to a full-on takeover, in the course of which the virus will induce them to reproduce its own DNA or RNA to make more virus. Smart thinking for mere chemicals.

CELLS THAT DON'T QUIT

Survival as a multicellular organism is one thing, but what about life at the cellular level? A sleeper hit in 2010, the book *The Immortal Life of Henrietta Lacks* documented the astonishing legacy of Lacks, who died of cervical cancer in 1951. Unknown to her—subsequently making for a difficult ethical position for her doctors—cells from her cancerous tumor had been taken while she was undergoing radiation treatment at Johns Hopkins Hospital and were subsequently used in lab research. One researcher, George Otto Gey, noticed how tough the Lacks cells seemed to be, outlasting all other cell samples, so he cultivated a line of cells from the originals, and named it the HeLa strain after Lacks. Nearly seventy years after her death, the HeLa strain is set to be truly immortal: still going strong, and used in laboratories all over the world.

45 WHY IS WATER SO IMPORTANT TO LIFE?

Water is needed by every living thing. From amoebas to elephants and everything in between, they've all got to have a drink. This is because water molecules are able to break down lots of different chemicals that living things need.

CHEMICAL SCISSORS

Water molecules are made of two hydrogen atoms and one oxygen atom in a triangle shape. The oxygen atom is larger and has more electrons around it. This means that a water molecule is slightly more negatively charged at the tip and thus more positively charged at the back. This quality is called dipolar. This charged state on water molecules makes it the perfect chemical scissors. The charged tip and hydrogen atoms are able to provide a strong force that can be used to break apart other molecules. This means that water is able to dissolve lots of different chemicals into the base components that are needed for all sorts of processes in the body, such as breathing.

OTHER USES

The ability of water to break down chemicals so blood can carry them around the body is incredibly important, but that's not its only purpose for living organisms. Water is central to many processes, from photosynthesis to respiration. Water is also able to help remove toxins from the body and, without it, any organism will dry up and no longer be able to take in the nutrients it needs, meaning that it dies. Water is therefore crucial to every single living organism. Water also forms a chemical bond called a "hydrogen bond," which is responsible for many of its useful properties. For example, hydrogen bonds cause lakes to freeze from the top down, creating a safe space for fish to live at the bottom during cold periods.

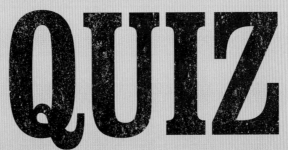

QUIZ

LIFE ON EARTH

Living things have existed on Earth for a great deal of time. Can you recall any facts from 100 million years ago?

QUESTIONS:

1. Did *T. rex* have more teeth than we do?

2. How many mass extinctions have there been on Earth

3. Ecologically speaking, where is wormy soil a bad thing?

4. What's special about the coelacanth?

5. How many soils does the USDA-created taxonomy define?

6. Which jellyfish defies the aging process?

7. How long does DNA last?

8. Apart from dating fossils, what is carbon dating useful for?

9. How much DNA do humans share with chimpanzees?

10. What does the IUCN Red List record?

Turn to page 246 for the answers.

BOTANY

46 WHAT'S THE WORLD'S MOST DANGEROUS PLANT?

Evolution has given plants all kinds of natural defenses against predators: poisonous chemicals to kick back with if they're eaten or sometimes even touched, and spikes, spines, and needles to repel attackers, to name a few. But if you lined them all up, which plant would win the most dangerous award?

LETHAL, TOXIC, AND MERELY IRRITATING

Lists of the "most poisonous" regularly put *Ricinus communis*, the castor oil plant, in the top spot (although it will be part of a very long list; many, many plants are toxic). The seeds are deadly: a tiny amount is enough to kill a human. Following closely behind is *Atropa belladonna,* deadly nightshade *(see left),* which, if eaten, will cause lurid hallucinations before it finishes you off. And then there's *Aconitum napellus*, monkshood, a powerful cardiac poison. None of these are believed to taste good, so there's no temptation to snack and you're unlikely to eat them by accident, although all have a history of being popular with poisoners. For direct fatalities, the world's biggest botanical killer is surprisingly obvious: it's *Nicotiana tabacum*, the leaves of which make smoking tobacco.

DEATH BY SHARK OR COCONUT?

Sometimes the damage a plant can do is purely accidental. For example, a coconut falling from height has the capacity to hurt—or even kill—someone standing below. How often does this happen? Probably more rarely than you might think. A favorite nugget that pops up annually in the tabloid press claims that you're much more likely to be killed by a coconut than a shark, supported by the "statistic" that 150 people are killed by falling coconuts every year. Traced back to source, this figure seems to have come from a bluntly titled paper, "Injuries Due to Falling Coconuts" published in the *Journal of Trauma* in 1984.

The science on which it is based seems shaky: deaths were based on the author's own experience treating coconut injuries while practicing in New Guinea, and appeared to have been arrived at by a simple multiplication of the estimated number of coconut palms growing globally. Furthermore, specifics on deaths-by-coconut weren't cited, only given as word-of-mouth stories.

This isn't to say that no one is ever killed by a coconut, rather that there's no way of knowing how many victims there are (though it's almost certainly fewer than 150 a year). And, going back to that tabloid headline, how likely are you to be killed by a shark? According to the International Shark Attack file, run by the University of Florida, there's an average of just six shark-related deaths globally per year. Unless you live unusually dangerously, you probably won't be killed by either.

A PRICKLY CUSTOMER

Prickly pear cacti, members of the *Opuntia* genus, won't make it onto any toxic lists, but gather the fruits unwisely and you'll find they rank very highly on the irritation scale. The plant is covered with highly visible and easily avoided spines; much harder to see are the superfine threadlike barbs called glochids. They "spring" off the plant easily, fasten into the skin, and are almost impossible to remove.

47 WHY ARE FOUR-LEAF CLOVERS RARE?

Rarities are always prized more highly than the commonplace, so it's natural that four-leaf clovers are the "lucky" ones, rather than the usual three-leaf kind. But they come about in the way of many other unusual occurrences—because of genetics.

IT'S IN THE GENES

Clovers, all 300-odd species of them, are unusual in that they are allotetraploid. This means that instead of coming in the more usual pairs, like those of humans, their chromosomes come in fours. In order to produce a leaf with four, not three, leaflets, all four of those chromosomes must be recessive. Even if the genetics are right, research shows that the plants also seem to need warmer conditions to help them produce the lucky variant. If everything is aligned correctly the plant may grow multiples, so if you do find a four-leaf clover, have a careful look at the other stems around it and you may find another. And if you're convinced that, for the proper luck of the Irish, what you actually need is a shamrock, don't worry—shamrocks and clovers are essentially one and the same.

Only around one in 10,000 plants has the genetic bias to four leaves. If you look for one methodically (as opposed to just coming across one by happenstance), the hunt may take you some time.

CAREFUL WHAT YOU WISH FOR

Clover isn't there just to yield the occasional helping of luck—it's also a crop plant that gives plenty of nutrition back to the soil and is widely used as animal feed. Most of the high level of protein it contains is found in the leaves, not the stems, so a plant that genetically favored the production of more leaflets would have extra commercial feed value, too.

OVERWHELMING LUCK

Clovers can come with even more than four leaflets. If you think that more leaves equals more luck, keep looking. The record breaker was a clover plant with a mind-boggling fifty-six leaflets. It was found by Shigeo Obara, a specialist in breeding and crossbreeding clovers, in Japan in 2009. It makes a mere four-leaf clover sound quite meager.

Korean studies in intensive clover breeding came up with increased numbers of leaflets in between 50 and 60 percent of some plants, but only when it had been exposed to radiation at the pollination stage. Meanwhile, experiments on its fellow crop, alfalfa (which also has three-leaflet leaves) have been made to try to get it to produce four or more leaflets regularly, but ultimately failed because the strain that was successfully developed not only grew extra leaves but also had much tougher and more indigestible stems. What the cattle gained in the leaves they were losing with the rest of the plant.

48 CAN TREES TALK?

Whether and how trees communicate has been intensively studied all over the world in the last twenty years, and the results have shown that, while they may not be able to talk, they have far more sophisticated ways of staying in touch than were previously dreamed of. To see their communication systems in action, you'd have to go underground.

TREE WHISPERERS

Two trailblazers, with years of study under their belts, have revolutionized our understanding of forest lifestyles. The first, Peter Wohlleben, a German forest manager and author, turned from commercial forestry to a much gentler form of tree husbandry when he began to understand how tree communities worked; the second, Suzanne Simard, a professor of forest ecology at the University of Vancouver, has spent many years conducting experiments into so-called mother trees and tree communities.

THE UNDERGROUND ECONOMY

There's a lot going on under the surface of the soil. Trees don't usually operate in isolation; they make extensive use of the mycorrhizal network, the vast systems of fine below-the-surface threads, mycelia, that are a largely unseen—though usually a much greater—part of the fungal life cycle. Tree roots and mycelia are symbiotic: their relationship has advantages for both sides. Mycelia work with trees in two ways: the first uses an ectomycorrhizal network, in which the fungal threads surround a tree root and work their way into it, in between its cells; and the second is by means of an endomycorrhizal network, in which the fungal threads actually pierce the cells and work inside them. Both trees and fungi enjoy several benefits. Trees draw phosphorus and nitrogen, nutrients that they aren't able to make themselves, from the mycelia, and pay them back with carbon, in the form of sugars.

ACTING IN BAD FAITH

If all this has left you with a comfortable picture of selfless trees acting in the common good, there's a more negative side to forest relations. Some species behave in ways that make it appear they're purely out for themselves. Black walnut trees, for example, produce toxins which they have been found to "share" with the trees around them, firing them through the mycorrhizal network. The benefit seems to be that they're rewarded with more resources, such as sunlight, because the trees around them won't intrude on their space.

COMMUNITY ACTION

The mycorrhizal network also helps trees to communicate with one another by sending chemical signals. One experiment conducted by Simard in 1997 involved her feeding trees with a traceable marker, then testing the trees around them. It turned out that the carbons had been carried to a large number of other trees, but those which had received the largest quantities were those that grew in shadier spots where they were less able to photosynthesize effectively. She concluded that the stronger "Mother" trees were sending help for the overall good of the group. What's more, she found that trees at the end of their life, either older or very sickly, would release their carbon stores into the underground system before they finally died, allowing them to be used by others.

49 WHY ARE PLANTS GREEN?

The wondrous bounty of nature is all around us in grasses, trees, and plants. There are a million varieties, but they almost all share a common element: green leaves. Plants are (probably) green because the Sun is green.

GREEN SUN

You might not think it, but the Sun is green. Like all stars, it gives off many different colors of light across the entire electromagnetic spectrum (even beyond visible), so the Sun is "white," as this is what you get when you mix all of the colors. However, the light that it gives off the most is green, so we would call the Sun green. It doesn't look green to us, because before the light reaches our eyes it passes through the atmosphere, which scatters the light, making it seem yellow.

PHOTOSYNTHESIS

Leaves are green because chlorophyll is green. Chlorophyll is the part of leaves that absorbs sunlight in order to get the plant the energy it needs for photosynthesis (which is how a plant creates its food). So the question, then, is why chlorophyll is green. Some people think that leaves are green because that way they can absorb the Sun's abundant green light. However, if the leaves are green, then that means they absorb everything but green, because green is the only type of light able to bounce off and into your eyes. If they really were adapted to absorb as much light as possible, then they'd be black. While there are some black plants out there, they are few and far between. The truth is, it's not really known why most plants are green, but the leading theory is that plants get their fill of sunlight without the green light, and if they took it in it would just cause them to heat up, which could be damaging. By being green, they can safely reflect the green light away.

FIVE QUICK FACTS

 1. "TREES" AREN'T A SCIENTIFIC CATEGORY

Trees don't make up a taxonomic group; they're species from a wide range of groups that have taken on qualities (extra height or spread, for example) that give them an advantage in their habitat.

 2. HOUSEPLANTS DON'T MAKE MUCH DIFFERENCE TO AIR QUALITY

Indoor plants might raise your mood, but when it comes to helping with air pollution, you'd need to grow at least ten large plants in every square yard of your home to reduce common air pollutants even a little.

 3. SEA SLUGS STEAL—AND THEN EXPLOIT—ALGAE

Animals eat, but plants don't need to; some species of sea slugs, however, have turned burglar: they consume algae then use the plant genes to photosynthesize for themselves, so they can make their own energy.

 4. THE LARGEST CARNIVOROUS PLANT IN THE WORLD IS *NEPENTHES RAJAH*

Nepenthes rajah, the giant pitcher plant, grows in Borneo and is able to trap (and digest) prey as big as a modest-sized rat.

 5. THREE-QUARTERS OF THE USES OF MEDICINAL PLANTS ARE LIINGUISTICALLY UNIQUE

Of 7,000 existing languages, 42 percent are endangered—mostly those spoken by small populations of indigenous people. If those languages become extinct, they will take an enormous body of botanical knowledge with them.

50 WILL GLOBAL WARMING MEAN THAT TREES GROW IN THE ARCTIC?

Global warming means that the range of many, many species is shifting. There are winners and losers in most categories, but how far can we expect that range to spread? Will trees end up growing high in the Arctic Circle?

RISING HEAT

For reasons that aren't completely understood, the rate of global warming in the far north is running at between two and three times faster than in the rest of the world. The northern treeline—the wavy line that literally runs around the top of the world at the point beyond which the climate is too harsh for trees to grow—is over 8,000 miles long, and north of it is the tundra, where the ground is often frozen just below the surface. Can the trees break through it as the climate warms?

SURVIVE, DON'T THRIVE

That's the positive argument. Naturalists who don't subscribe to it argue that even some distance south of the current treeline, the conditions for plants are harsh, and those trees that succeed in living there grow very, very slowly. A study carried out in Alaska in 2016 found that a tiny spruce no taller than a person was nearly a century old. Both nutrients and light are in short supply, and it's easier for weaker saplings to be shaded out by marginally more successful ones. Trees near the treeline may survive, but they couldn't be said to be thriving.

Other arguments point out that species other than trees would also be likely to take advantage of warmer surroundings. Invertebrate species could increase their range, and these might include pests that would live on—and weaken—the trees. Higher temperatures would raise the likelihood of wildfires, and ultimately, if the heat continues to rise, species that were previously happy may find that their new habitat has become too hot.

FOSSIL LESSONS

Trees have spread into territories previously too cold for them many times before. Around 21,000 years ago, toward the end of the last ice age, trees in Europe were limited to present-day Italy, Greece, and southern parts of Spain. But as the northern lands gradually thawed, the pollen record shows the way in which the trees advanced. By human standards it was slow—it took a mere 18 millennia for tree cover to move north from central Europe up to present-day Britain and Scandinavia.

MARCHING NORTH

At the moment, experts' views are divided. To extend the spread of their habitat, trees have to produce seeds and some must move into new territory, usually by being taken up (by insects, birds, animals, or the wind) and set down in a new place to grow. They have to germinate, survive the vulnerable seedling and sapling stage, and make it to adulthood, at which point they can set seeds themselves and continue the habitat spread. The research modeling programs that have been used to analyze data have predicted the likely rises in temperature in the Arctic over the next few decades will result in tree cover marching north of its current line, displacing smaller plants and shrubs. Meanwhile shrubs would push farther north, preparing the way for even more trees in the future.

52 HOW OLD IS THE OLDEST TREE?

Many trees far outlive the human life span, but how old can a tree get (and which is the oldest tree in the world)? Trees have been around for a long time—the earliest identified species are in fossil form, dating back to the late Devonian era, 385 million years ago. Many of today's species are pretty ancient, both in individual and group terms.

AS OLD AS METHUSELAH

The bristlecone pine (*Pinus longaeva*), an evergreen that lives in the western United States, seems to operate in slow motion. Not only do individual trees have life spans counted in thousands rather than hundreds of years, but even the needles last longer on the tree than those of any other species, staying on their branches for up to three decades. One example, growing in the Inyo National Forest in California, is known as Methuselah and held the title of the world's oldest tree for some years. It was aged at 4,845 years in 2013—but that same year it was knocked off its pedestal when another of the same species was discovered to be 5,062 years old.

COUNTING IN MILLENNIA

Methuselah and friends are mere youngsters, however, if you include clonal trees in the count. Clonal trees are tree groups that are connected by their root systems, meaning that each individual is genetically identical to others in the group—so their age is judged to be that of the colony overall. Probably the best known is the Pando, a huge clonal group of quaking aspen trees (*Populus tremuloides*) in Fishlake National Forest in Utah, which is believed to have been there for a jaw-dropping 80,000 years. There's a third "oldest" category, for the world's oldest individual clonal tree—and the top place in that is currently held by Old Tjikko, a Norwegian spruce (*Picea abies*) that survives on a windswept Swedish mountainside. Spruces can generate new roots from their branches, which is probably why Old Tjikko has managed to clock up an impressive age of 9,565 years—and counting.

HOW CAN YOU TELL?

It's all very well to claim thousands of years' worth of birthdays for a tree, but how can you tell? The traditional, and best-known, method is dendrochronology—counting the age rings in a cross-section of the trunk. The trouble is, to use it, you need to cut down the tree first. Alternatively, you can use a tool called an increment borer to push into the center of the tree, then remove a very fine cylinder of wood, which causes minimal damage and will show all the age rings in a less destructive way. The most ancient trees are usually carbon-dated for an accurate (and often impressive) result.

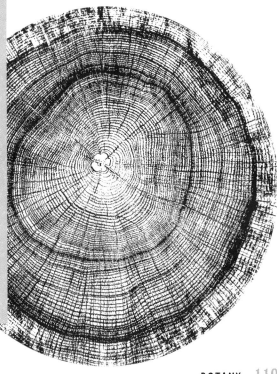

53 WHICH TREES WORK BETTER—THOSE WITH LEAVES OR NEEDLES?

Deciduous trees grow and drop leaves every year, and the whole process takes a huge investment of energy. On the other hand, trees with longer-lived needles rather than leaves may not be able to photosynthesize as efficiently. What are the deciding factors on which works best for different species?

BALANCING THE BOOKS

Actually, photosynthesis has less to do with the question than you might think: although in most cases needles will stay on a tree year-round, they still won't be doing much photosynthesis in the darker, colder months of the year. But needles have other advantages when it comes to conserving a tree's resources. They're structured very tightly, with a central vein and far fewer stomata, or pores, than the leaves of broadleaf species, and they're coated with a tough, resistant cuticle that waterproofs each needle. In climates that are very dry or very cold, or both, they are effective in cutting water loss to a minimum— important in locations where rain is rare, or where the ground can freeze, which will stop a tree from taking up water through its roots. They even contain built-in antifreeze in the form of cryoprotectants, which lower the freezing temperature of the water in the tree's cells and helps to keep their needles from freezing in even the coldest weather.

Needles are economical, too: by staying on the tree for several seasons, they help it recoup its initial investment by spreading it across two or more years. There are a few exceptions to the needles-equals-evergreen rule, which do shed their needles annually; they tend to be species that have to survive extremely tough conditions, so it's probable that, for them, even the comparatively low cost of maintaining needles is too big a price to pay over a long, harsh winter.

SURVIVING SNOW

Evergreens have one other advantage over broadleaf trees in areas that experience heavy snowfall. The narrow needles usually combine with a drooping habit overall, with trees' branches bowed slightly downward. As a result, snow that might break branches and damage a more rigid, upstanding shape simply slides off evergreens onto the ground.

IT'S NOT EASY BEING GREEN

Deciduous trees have a budget, but it works in a rather different way. In general, they prosper in more moderate climates than evergreens and, having put the energy into growing leaves that will last only a few months, they make the most of their investment: deciduous leaves have plenty of pores and are structured to maximize photosynthesis whenever light levels allow. During leaf drop, in the fall, the trees won't discard any content that can be recycled. As the leaves change color, they turn from green to yellow and orange, which signals that the mother tree is withdrawing the chlorophyll from the leaves back into the trunk. It's the chlorophyll that absorbs sunlight and helps to power photosynthesis, and the tree won't waste it; it'll be stored and saved for next year.

ARE VITAMINS GOOD FOR THEIR PARENT PLANTS?

We're all used to the idea that eating plenty of fruits and vegetables will give us most of the vitamins we need to stay healthy but that our systems won't produce on their own. But do plants use vitamins in the same way that we do?

PEPPING PLANTS UP

Research into the ways in which plants use vitamins for themselves, as opposed to looking at the benefits vitamin-rich plants offer to those who eat them, is still at quite an early stage, but three specific vitamins have been studied: C, E, and B. Vitamin C turns out to be just as important for plants as it is for people—they need it to photosynthesize, and it also helps them to withstand the harmful effects of ozone (the latter can enter the plant's stomata, porelike openings in its leaves, and burn them from the inside). Vitamin E is less crucial, but plays a beneficial role in protecting plants against low temperatures, helping them to withstand cold wilt and acting as a sort of built-in antifreeze. How plants use their vitamin B hasn't been conclusively decided—studies gave mixed results, with some showing that vitamin B helps plants get and stay more robust and encourages root development, and other studies didn't seem to identify much impact on their health.

MAKE YOUR OWN

Not only do plants make their own vitamin C, but they also manufacture it in the quantity they need, adjusting it according to their circumstances. Because it's a powerful antioxidant, it can protect them from oxidation damage if they're enduring intense levels of light or living through drought conditions.

QUIZ

BOTANY

From plants that smell like meat to communicating fungus, the plant world is full of wonders. Can you remember much about them?

QUESTIONS:

1. When do the earliest tree species date from?

2. What's the highest number of leaflets ever found on a clover plant?

3. Apart from the smell of decomposing flesh, what other tactic does the titan lily use to attract beetle and fly pollinators?

4. What is the world's heaviest flower?

5. How long do the needles of the bristlecone pine last on the tree?

6. What are clonal trees?

7. How does the waxy coating on tree needles help the tree?

8. What is deceptive pollination?

9. What are glochids and what do they do?

10. Why don't black walnut trees make good neighbors?

Turn to page 246 for the answers.

ZOOLOGY

WHY DO FROGS BLINK WHEN THEY SWALLOW?

If you've ever watched a frog eating a fly, you'll have noticed that it will briefly close its eyes as it swallows a mouthful. (Sometimes it also uses its front legs to bundle its prey into a manageable shape before stuffing it into its mouth.) The overall effect is rather charming: it looks like a small gourmand really relishing its meal. What's really going on, though, is purely practical.

CHOWING DOWN

When you eat, first you chew and then you use your muscular tongue to help you swallow. Many frog species, on the other hand, don't have teeth, and the teeth of those that do grow in the upper jaw only and are used to help the frog to grab and secure its food rather than to chew it. Frogs' tongues are also attached to the front rather than the back of the mouth, so aren't terribly useful when it comes to swallowing action. As a result, frogs need an altogether different technique to swallow. Once the prey is in its mouth, the frog closes its eyes and its large eyeballs retract slightly into its head, causing them to press down on the roof of the mouth and push the food back toward the gullet, enabling the frog to gulp it down. From the early twentieth century, it was widely accepted by scientists that frogs swallow their food using their eyes; however, in 2004 this was tested under rigorous

scientific conditions. In the study, carried out by the University of Massachusetts, northern leopard frogs (*Rana pipiens*) were examined while eating, and it was found that if a frog was stopped from blinking as it ate, it had to swallow more times to get the mouthful down—2.4 times with eye retraction went up to four gulps without.

FROG VS. FLY

If all this eye action makes eating sound rather too much like hard work, frogs more than compensate when it comes to their tongues. Frogs' tongues are around a third of the length of their bodies and have some extraordinary qualities. They're fast—as they need to be when their owners are pursuing flies—and they're also extremely sticky, covered with many hundreds of glands that secrete a sticky saliva. The saliva is able to change texture during the hunting process; in the mouth, it's thick and gloopy, but as the frog's tongue shoots out toward an insect, the saliva liquefies and becomes runny, quickly coating the prey as it comes into contact with it. Once the prey is secured, the saliva around it thickens again, immobilizing it and making it easy for the frog's tongue to reel the meal back in. The whole textural-saliva trick—thick to thin and back again—takes just a fraction of a second.

Bon appétit! We hope we haven't put you off your lunch.

HOW CAN VIRUSES JUMP BETWEEN SPECIES?

Since the arrival of COVID-19 in 2020, this question carries an urgency that many people may not have felt before. Just how often do viruses jump between species, and how do they do it?

IT'S COMPLICATED

Given how many unknown viruses are believed to exist (an estimated 1.7 million across mammals and birds alone), they cause problems relatively rarely. A virus that successfully moves from one species to another has made what's called a zoonotic jump. And it's likely that the majority aren't ever identified—either because the attempted jump was ultimately unsuccessful or because the effects of the jump aren't very significant, and so the fact that they're due to a new (cross-species) virus isn't ever noticed. Something with the extraordinary "success" rate of COVID-19, both readily transmitted and causing serious illness in a high number of subjects, is a very rare virus indeed.

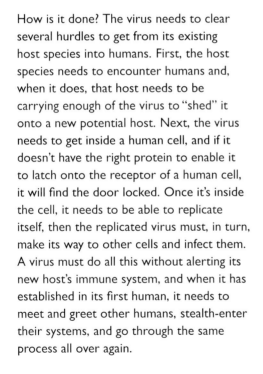

How is it done? The virus needs to clear several hurdles to get from its existing host species into humans. First, the host species needs to encounter humans and, when it does, that host needs to be carrying enough of the virus to "shed" it onto a new potential host. Next, the virus needs to get inside a human cell, and if it doesn't have the right protein to enable it to latch onto the receptor of a human cell, it will find the door locked. Once it's inside the cell, it needs to be able to replicate itself, then the replicated virus must, in turn, make its way to other cells and infect them. A virus must do all this without alerting its new host's immune system, and when it has established in its first human, it needs to meet and greet other humans, stealth-enter their systems, and go through the same process all over again.

GOOD VIRUS, BAD VIRUS

It's easy, looking at COVID, rabies, hepatitis, and influenza, to name just a few, to decide that viruses aren't good for humans. But they have an essential role to play in natural balance. Tony Goldberg, an epidemiologist at the University of Wisconsin–Madison, puts the position bluntly: "If all viruses suddenly disappeared, the world would be a wonderful place for about a day and a half, then we'd all die." Why? Because viruses exist at many different levels of life and are an essential component in controlling and balancing the environment. Viruses of bacteria, called phages, for example, kill off over-competitive bacteria, benefiting other populations. Kill off all viruses and biodiversity would likely reduce very fast— probably with catastrophic results.

57 WHY CAN'T AN ELEPHANT BE THE SIZE OF A HOUSE?

By animal standards, elephants are big: a full-grown male can weigh in at over 7 tons. But even the largest elephant is a lot smaller than a blue whale, although it's also an awful lot bigger than the largest spider. What limits the size of elephants (and spiders)?

THE *RIGHT* SIZE

Nature is nothing if not practical; most animals are the optimum size for the lives they lead, or they would die out. The laws of allometry—the science of biological scaling—dictate that an animal that increases in mass by a certain amount will need a commensurately greater increase in strength. An outsized elephant would need a massively thickened skeletal structure to manage its increased bulk, and it would quickly become impractical for it to move around, to feed itself, and to reproduce. Back in 1928, in an essay called "On Being the Right Size," the scientist J. B. S. Haldane pointed out that a fairy-tale giant who was 60 feet tall but human-scaled would break its puny leg bones every time it tried to walk.

Put a mammal in the sea, and gravity, one of the main limits to growth, is removed. With water to bear its weight, the blue whale, the largest animal on Earth, can reach a length of 105 feet and a weight of 165 tons.

SPIDEY SENSE

At the other end of the scale (arachnophobes might want to stop reading now), hugely enlarged insects and spiders, which are quite differently designed from mammals, would face several issues. A spider's rigid exoskeleton would become very cumbersome if it were to grow much heftier than that of the largest species (for the curious, that's probably the Goliath birdeater which, weighing in around 7 ounces, and about 12 inches long, is quite big enough for most people). Unlike mammals, spiders have to undergo successive molts as they mature, and there's a brief period in each molt when, unprotected, they are particularly popular with predators. More growth could mean more molts, and increased periods of vulnerability. Not only that, but spiders also have

relatively basic breathing apparatus; most take in oxygen passively through trachea, pipes located along their sides, and this simple system would be inadequate if it needed to feed oxygen around a spider with a much larger mass.

DINOSAURS: A ROARING SUCCESS?

If the size of an elephant is scientifically fixed, how come dinosaurs could grow so much bigger? One major factor is that they were reptiles, not mammals—being cold-blooded, they needed much less energy to keep their metabolisms on track. Prehistoric times also produced some huge insects, much larger than any found today. (Remember the vast dragonflies flying around in *Jurassic Park*? Unlike some aspects of the movie franchise, their existence is supported by the fossil record.) They may have evolved successfully because all those millions of years ago, the Earth's atmosphere had much higher oxygen levels (around 35 percent to today's 21 percent), which could have enabled a scaled-up version of insects' comparatively rudimentary breathing systems to work.

58 WHAT'S THE FASTEST ANIMAL IN THE WORLD?

If you spent any time reading record-breaker lists as a kid, it's probably still embedded in your mind that the fastest animal on Earth is the cheetah (*Acinonyx jubatus*), which can reach a land speed of 0 to 60 mph in under 3 seconds, and has been recorded, briefly, hitting 75 mph.

That's certainly fast, but in the air it's far outpaced by the hunting dive of a peregrine falcon (*Falco peregrinus*), which, though short, can hit an extraordinary 240 mph. Its regular flapping-along speed, on the other hand, is a comparatively leisurely 60 mph, and here it is surpassed by the small but speedy Mexican free-tailed bat (*Tadarida brasiliensis*). The latter was considered a neat but not outstanding flyer until tests carried out in 2009 on a bat loaded up with a transmitter recorded it attaining short bursts of flight at an astounding 100 mph.

NOT SO FAST

That's the earth and the air; in the water, the fastest swimmer recorded is the sailfish (*Istiophorus*) at 68 mph, with swordfish, tuna, and pilot whales following not far behind.

Of course, these aren't the only animals to be quick on their feet (or paws, or fins)—only the outstanding ones. However, if you start to look at the fastest movement made by an animal, you shift into a much smaller and less showy category of record-breakers who deserve their own fifteen minutes of fame.

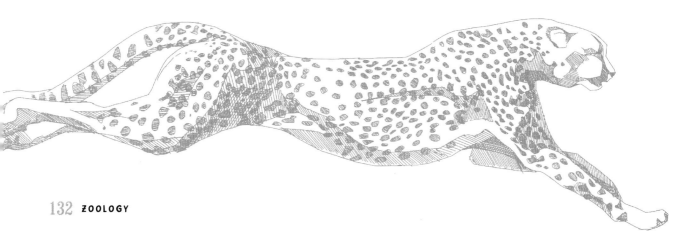

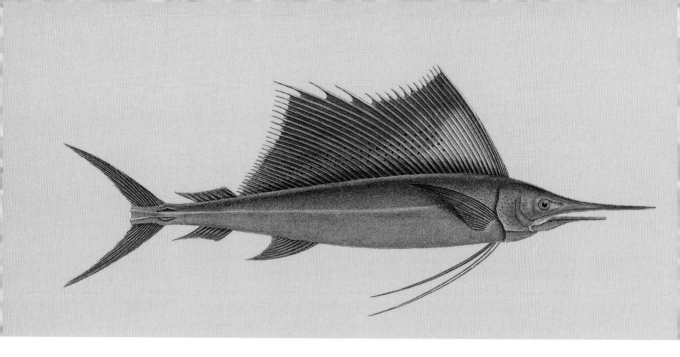

SNAPPY MOVER

For several years, the record for the fastest movement was held by the punch of the mantis shrimp (*Stromatopoda*). Although the shrimp itself is only 4 to 8 inches long, its upper arm has a spring-loading structure, the saddle, above the club at its "hand" end, which gives it an outstandingly powerful punch. Muscles at the end of the saddle pull on it, arching it like a bow, so that when it's released, the club can shoot out for the knockout blow (51 mph), which easily demolishes the tough shells of the snails, small crayfish, and so on which are a large part of the shrimp's diet.

The mantis shrimp's record was broken in 2018, although not before its structure had been the model for a tiny "punching" mechanism which, in time, may be used to build power into new robotics. Who now holds the crown for fastest movement? That would be the Dracula ant (*Mystrium camillae*), whose mandible snap goes from 0 to 200 mph in 0.000015 seconds. As was widely reported at the time by ecstatic researchers, who used an ultra-high-speed camera to capture the feat, that's five times faster than a human blink. Again, the secret is in spring-loading: the ant prepares its mandibles before the snap, repeatedly pushing their tips together until they are fully charged with energy, then releases them with a force guaranteed to knock out its opponent, whether enemy or prey.

59 WHY AREN'T MORE ANIMALS COLORED BLUE?

Blue does happen in nature—there are blue flowers, butterflies, birds, and more—but generally it's found much more rarely than other colors.

A TRICK OF THE LIGHT

With its short color wavelength, blue light is more easily absorbed than any other. Light waves fall on a spectrum, and when we're looking at something, the color it appears to be is that which reflects most strongly off it. The longer the color's wavelength, the more visible the color, and red and yellow have longer wavelengths than blue.

When you do see blue, it's because of structural color—in other words, the plant or animal has developed a way to cheat the short wavelength. Some species of morpho butterflies, found in Central and South America, have wings of the most brilliant blue you'll ever see. Looked at under the microscope, this turns out to be due to a very specific structure in their wing cells, which deflects the light as it hits them and reflects only the blue. Birds that appear blue have feathers that are structurally adapted in a different way: the cells of their wings have a complex structure made up of keratin and air pockets, which, when light hits them, cause the red and yellow to cancel each other out, while enhancing the blue. Whether you're looking at a bluebird or a peacock, that vivid shade is literally a trick of the light.

YOU ARE WHAT YOU EAT

Couldn't animals turn themselves blue by eating blue food? Not easily. Color can be affected by diet to some extent; flamingos, for instance, are born gray, turning pink only when they've eaten enough of the shrimp that are their natural diet. Blue foods, though, are in very short supply.

FIVE QUICK FACTS

 1 SWIFT BY NAME

Over its lifetime, a swift can fly over 770,000 miles—that's the equivalent of more than thirty times around the Earth, or three trips to the Moon.

 2 FLIES LIVE LIFE IN SLOW MOTION

It's hard to swat a fly because it experiences time more slowly than we do, allowing it to take leisurely evasive action. To a fly, a second of human time feels like four, because the fly's "flicker fusion rate" is more than four times faster than a human's.

 3 SEA OTTERS HAVE THE THICKEST FUR IN THE WORLD

Their skin has up to a million hair follicles per square inch. Unlike many other sea-dwelling mammals, they're missing a layer of blubber, so they need their extravagantly warm coats.

 4 NOT ALL SPECIES BLEED RED

Blood color depends on the proteins and minerals it contains. Your blood is red (because the oxygen in your blood is carried by hemoglobin, which is rich in iron), but an octopus bleeds blue (because it uses hemocyanin—which is rich in copper—to do the same job).

 5 KOALAS SPEND TWENTY-TWO HOURS PER DAY ASLEEP

Given its name, you might think that the sloth would win gold in a sleeping contest. Wrong! Koalas average twenty-two hours asleep in every twenty-four, against the mere twenty hours of sleep notched up by the sloth.

60 WHO HAS THE BEST ECHOLOCATION?

When it comes to precise navigation, the animal who takes gold may be a surprise. It's the narwhal (*Monodon monoceros*), a member of the toothed whale group. It's easily recognized because of its long, twisted, "unicorn" horn, and it also has the finest echolocation skills in the ocean.

CLICKBAIT

Toothed whales and dolphins navigate with clicks. They make the noise, and the reverberations that come back from the noise help them to create a picture of their surroundings, and how they're placed within them. Narwhals use their echolocation to manage a very difficult environment. They live high above the Arctic Circle, under water which is itself under a thick layer of ice for much of the year, coming up for air in small areas of open water, known as polynyas. A geophysicist from Hokkaido University, Evgeny Podolskiy, described the ocean the narwhals swim in as a very loud place indeed: "There is so much cracking due to ice fracturing and bubbles melting . . . it's like a fizzy drink underwater." In 2013, a study was made of the narwhal population of Baffin Bay, west of Greenland; it was found that to cope with the tough environmental challenges, the animals could produce a record-breaking thousand clicks per minute.

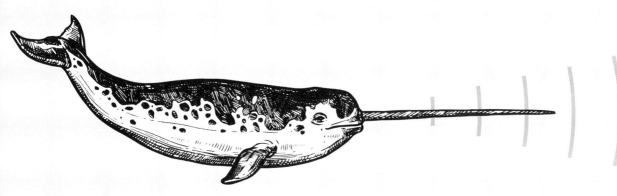

HOW IT'S DONE

The noises that the narwhal produce are made in its nasal passages, then a fatty structure at the top front of its head, called a melon, gathers them into a "beam," which it sends out into its surroundings. When the beam hits something, the echo bounces back and hits another fatty structure in the lower jaw, which generates a reading. The narwhal can narrow the beam as well as casting it over a broader area, rather as you might cast a general glance over a landscape, then narrow your focus to the specific part of it that interests you.

As well as its echolocation skills, all male narwhals—and a small percentage of females—have the distinctive twisted horn. Technically, because it's the animal's modified front left tooth, it's a tusk. It's around 10 feet long and highly sensitive, with a hard inner structure and a softer outer layer, carrying ten million nerve endings. These help its owner to sense temperature, water pressure, and salinity, so the horn has practical uses (although it's believed that it also plays a part in mating displays).

AIRBORNE CLICKS

Marine mammals are far from the only animals to use echolocation. Bats use sonar to navigate, but through air rather than water (and their highest click rate is around 200 per minute, just a fifth of the rate of a narwhal). Insectivorous bats hunt in the pitch dark, which doesn't stop them from being highly effective, catching up to half their body weight in luckless moths and other insect prey in a single night. In defense, some species of moths have adapted to produce their own clicking noises, which jam a bat's sonar just long enough for the moth to get away.

61 ARE THERE MORE COWS THAN PEOPLE?

According to a study published by *The Economist* in 2011, there were around 7.7 billion people and only 1.5 billion cows in the world, so people were outnumbering cattle by more than five to one. However, add in a billion sheep, another billion pigs, and 19 billion chickens on the planet and the balance changes sharply. Overall, Earth has far more farmed animals than humans, and in some countries they substantially outnumber people—Brunei, for example, clocks up around 40 chickens per human inhabitant.

THE RISE OF THE CARNIVORE

If you were only reading parts of the Western press, you might have the impression that vegetarianism and veganism are on the rise, and in some countries they are. Globally, however, we're actually eating three times as much meat as we did fifty years ago, against a mere doubling of the world's population. Where people are eating meat, however, has changed somewhat: European meat consumption has leveled off (or in some areas declined slightly), and North America's has risen a little. Both were originally at the highest levels in the world, but what's made a huge difference to the global picture is the appetite for meat in countries that previously ate little: China eats more than twice as much meat as it did three decades ago, and the rest of Asia and Central and South America show rises, albeit smaller ones. African consumption is rising slowly, from a low initial figure. With a population that has a high percentage of both vegetarians and vegans, only India's appetite for meat remains both steady and low.

WILD DISCREPANCIES

While farmed animals outnumber humans, figures published at the end of 2020 showed that numbers of wild animals are plunging. Humans represent 36 percent of the world's living mammals, but mammals in the wild represent just 4 percent. At a staggering 60 percent, farmed mammals make up the difference.

SACRED LIFE

An estimated 37 percent of India's population is vegetarian—against around 5 percent in Australia and 9 percent in Germany—and a fair percentage is vegan. When it comes to the diet with the most firmly held views on life, that prescribed by Jainism is probably the strictest. Jainism is an ancient religion with a strongly held tenet that all living things have souls. While this means that dairy foods are allowed, vegetables pulled whole from the ground, such as onions, garlic, or carrots, aren't, because once pulled up they are unable to regenerate themselves. Jain monks and nuns also wear masks to ensure they won't accidentally swallow even the smallest insect, and use small brooms to sweep the ground before them as they walk to avoid harming tiny creatures in their path.

62 IS THERE A WAY TO SPEED UP EVOLUTION?

Evolution is notoriously a slow business. It's believed to be made up from a mixture of the Darwinian concept of survival of the fittest, and the effects of chance happenings on different species populations, although there's still a range of opinions on exactly how it works (and just how fast—or slow—it is). One thing that experts agree on is that evolution never stops. And there may even be some ways of making it go faster.

HOW LONG DOES IT TAKE TO MAKE A NEW SPECIES?

Around a million years seems to be the historical answer. Evolution happens when genes mutate and the mutation persists through many generations until it becomes the norm. When individuals with specific mutations not only survive but thrive and breed in a particular environment, their offspring inherit those mutations, which may give them a built-in advantage—the survival of the fittest. Sufficient changes ultimately mean the creation of a new species.

But changes don't always persist. A wide-ranging study made by a team of researchers from the Universities of Oslo, Norway, and Pretoria, South Africa, and published in 2011 combined what was known about short-term changes

that might happen over just a few hundred years with accumulated knowledge about long-term evolution—changes that could be traced far back, often across millions of years. It concluded that many short-term changes don't last the course all the way to the creation of something altogether new.

SPECIES POCKETS

They may, for example, be limited to a small "pocket" of a species, or they may fall victim to environmental changes that outpace them, or altogether random events may interrupt and disrupt them. Despite the evidence of plenty of shorter-term changes, the estimate that it takes around a million years to create an entirely new species remained unchallenged.

QUICK FIXES

In this shorter term, there's evidence that some human genes have adapted much faster than might be expected. Over the last four millennia (a long time in human terms, but barely a dot on the long line of evolutionary changes), specific diets and living circumstances, as well as resistance to some diseases, all seem to have resulted in genetic changes to some groups in some places. For example, the Inuit appear to have adjusted to the high-fat diet necessitated by their environment, while people living at high altitude—populations in the Andes or the Himalayas—can deal with lower oxygen levels than would be managed by those living at lower altitudes. There's even some evidence that humans in Western societies may be adjusting genetically to cope better with today's often high-sugar, high-fat diet. However, it will probably take many more millennia before we know if those adjustments are to form part of long-term genetic changes.

63 HOW SMART ARE SNAKES?

In a world that is judged by humans, how clever are reptiles? Historically, intelligence in others has been measured by how close it is to our own, and the same species top animal intelligence charts over and over again: chimpanzees, dolphins, and elephants are always winners. Corvids and parrots rank highly—bird intelligence has also been the focus of plenty of experiments. And cephalopods, in particular octopuses, have become famous for being smart (and for their escapology skills), often outwitting those who are supposed to be testing them.

COLD-BLOODED COGNITION

Reptiles have been the poor relation in intelligence tests, credited mostly with instinctive, hard-wired behavior without much capacity for learning and adapting. However, looked at again in recent decades, it turned out that the more likely problem was with the tests, many of which were inappropriate to reptile abilities. Not only do reptiles need a warmer environment before they become active, they're also not necessarily motivated by the food rewards used with most other animal groups, as they can be very irregular eaters. Exposed to aversive stimuli, their reaction tends to be to freeze rather than move. In short, they're neither mammals nor birds nor cephalopods, and they need their own tests.

In more sympathetic experiments, many reptiles showed clear evidence of learning. Indigo snakes learned to operate levers, and Burmese pythons to press buttons. Studies done in reptile habitats saw cave-dwelling Cuban boas forming packs to hunt bats. It's possible that some reptiles may even enjoy play: tortoises, turtles, and some lizards have been recorded manipulating objects apparently just for fun. Komodo dragons, the gigantic monitor lizards from Indonesia, have been known to indulge in tug games with their keepers—although it's not entirely certain what their intentions are.

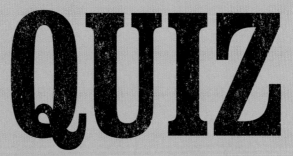

QUIZ

ZOOLOGY

Are you as smart as a snake? As clever as a Cuban boa? Take this zoology quiz and find out!

QUESTIONS:

1. What's special about frogs' saliva?

2. How many viruses are believed to exist in mammals and birds?

3. Why can a sea-dwelling mammal grow bigger than a land-dwelling one?

4. In the Jurassic era, how much oxygen was there in the atmosphere?

5. Why won't followers of the Jain religion eat onions or garlic?

6. How have human populations living at high altitudes evolved quicker than expected?

7. How do some moths escape from bats?

8. What has the narwhal's tusk evolved from?

9. How fast can a peregrine falcon dive when it's hunting?

10. Which domesticated farm species far outnumber people?

Turn to page 247 for the answers.

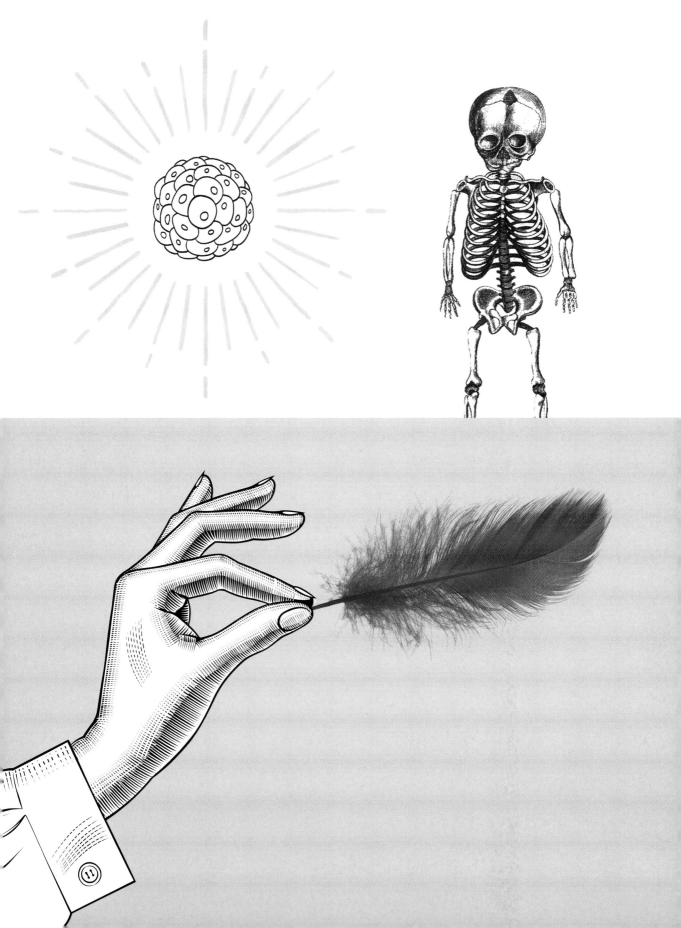

THE HUMAN BODY

64 WHY DO BABIES HAVE SO MANY BONES?

They may be small, but babies notch up a total of
300 bones apiece—considerably more than adults do.
By the time they grow up, the total will have reduced:
grown-up humans have a comparatively modest
206 bones. What happens to the extras?

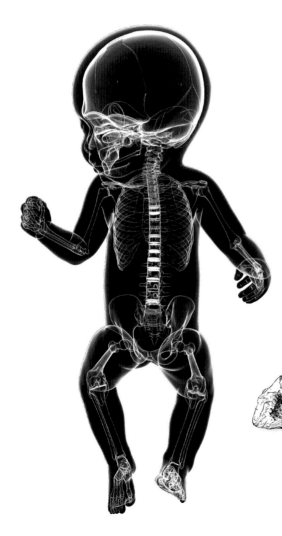

BONE VS. CARTILAGE

The effort of the human birth process
has been described as being like trying to
squeeze a grapefruit out of your nostril—
and humans have the most protracted
births of any mammal. In order for it to be
actually possible for the mother to push the
baby through the birth canal, the newborn
needs to be very bendy indeed. So, instead
of rigid bone, a large number of those
300 baby "bones" are made up of cartilage,
which is a much softer and more elastic
substance. As an adult, pads of cartilage
will remain at the ends of your long bones,
cushioning the joints, as well as in your ribs,
ears, and nose, but most of it will have
fused into fewer, larger pieces of bone.

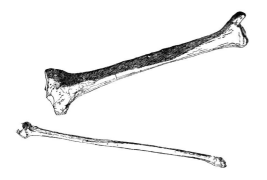

HARDENING-UP

The technical name for this hardening-up is endochondral ossification. Cells called osteoblasts build bone, gradually replacing the cartilage from the inside out. The baby cartilage "skeleton" offers the template on which bone will grow, but the cartilage itself doesn't become bone.

The key element the osteoblasts need for the process is plenty of calcium and vitamin D, which aids the absorption of calcium. This is why so much emphasis is placed on calcium intake for pregnant women and nursing mothers. After birth, calcium is passed on to the infant in breast milk, and babies and children, unlike adults, can "bank" calcium stores.

HOW LONG DOES IT TAKE?

Babies stay bendy for quite a while. If you've ever marveled at the resilience of a child at the crawling-to-walking stage, and its ability to fall constantly without hurting itself, the reason is its retained flexibility. The solid-bone skeleton doesn't form all at once; the growing of bones is a long process that isn't completed until a person hits their early twenties. And bones change even in adulthood—bone can grow to support and mend injuries. Your skeleton also "maintains" itself, replacing upward of 5 percent of its bulk with fresh bone every year.

BONE-HEADED

Most people have felt the softness of baby bones for themselves when they've touched the area around the anterior fontanelle, or soft spot, at the top of a baby's head. This is the most obvious sign that the five overlapping plates that form the baby's skull haven't fused yet. There's also a second soft spot, the posterior fontanelle, at the back of the skull. Although the fontanelles are both closed by the age of three, the skull's bony plates are held together by flexible structures called sutures, and won't finally fuse into a single structure until the brain has reached its full size, when the child is around seven years old.

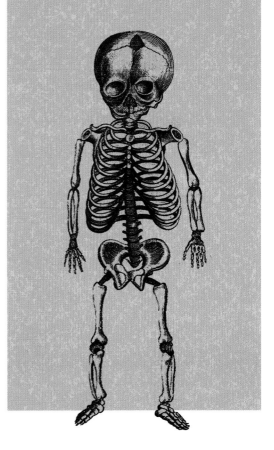

65 WHY ARE HUMANS BIGGER THAN THEY WERE 500 YEARS AGO?

It's common knowledge that humans are much larger and taller than they used to be. Like all "common knowledge," though, this doesn't tell the whole story. Our species' growth pattern is a story of fits and starts, rather than a steady progression.

HOW BIG DID WE USED TO BE?

Looking at ancient suits of armor, you might conclude that humans have become a lot bigger. After all, there's not much chance of wriggling into a suit made from metal—it has to fit, and some look very small. But although humans are generally taller than they have been at some points in the past, the rate at which this change has occurred has been far from steady. Scientists have found that the size increase has happened erratically, and most growth graphs have some notable blips along the way.

UPS AND DOWNS

Most accessible long-term studies deal with America and northern Europe, so they're necessarily limited. One, published in 2017 by the University of Oxford, looked at the average height of men in England over a 2,000-year period. Its findings revealed that height went down in some periods as well

DEM BONES

How do you assess the height of someone when you have only their skeleton to go by? Forensics teams measure a long bone, usually the femur or thighbone (although the long bone of the arm, the humerus, is also used), then apply an equation—for example, measuring the length of the femur in centimeters, then multiplying the measurement by 2.6 and adding 65 to get the height. Calculations vary a little according to the race and gender of the skeleton's original owner.

upward growth has come over the last two centuries; from the middle of the nineteenth century, northern Europeans and Americans have become steadily taller, with North Americans holding the record as the tallest people in the world over the period between the American Revolution and the end of World War II. Since then, they've been overtaken by a number of European countries, with the Dutch currently holding the record; the average Dutchman measures just over 6 feet tall.

NAPOLEON COMPLEX

Final food for thought: The Napoleon complex, said to afflict short men, is a modern misnomer. In fact, Napoleon measured 5 feet 6 inches—well above the average (around 5 feet 4 inches) for his time.

as up, the determining factors including long-term weather conditions (for example, the "Little Ice Age" that began in the fourteenth century), periods of prosperity (ensuring a good diet, particularly important in childhood), and the different lifestyles of town and country dwellers. But both this and other broader studies had some surprises.

A 200-YEAR GROWTH SPURT

Between the end of the medieval period and the start of the eighteenth century, people actually got shorter—by 1700, the average northern European man was a full 2.5 inches smaller than he had been back in the eleventh century. The most consistent

66 HOW MANY TYPES OF BACTERIA LURK IN YOUR GUT?

We hear a lot about the importance of the microbiome in your gut. The last decade has seen an explosion of interest in the workings of the human digestive system, and new studies have made the bacteria in your colon answerable for everything, from stress levels to intelligence.

GUT INTERESTS

Your gut holds trillions of bacteria that weighs an estimated 2 to 4 pounds per person. They are highly individual; most people harbor between 500 and 1,000 types, both beneficial and not so good, and everyone has a slightly different formula.

DO YOU HAVE A GUT TYPE?

Most people know their blood type, but do you know what type of gut you have? In 2011, a study in Heidelberg, Germany, found that there are three broad profiles of guts with identical microbiomes, as they shared most of the same bacterial groups and missed others altogether. The people who shared each profile seemed to have nothing else specifically in common and were a mix of genders, ages, weights, ethnicities, and states of health.

THE GOOD, BAD, AND USEFUL

Scientists hope that in the future it will be possible to tailor medical treatment according to the "formula" of a patient's gut, adding microbial reinforcements to the bacteria already there. With antibiotic resistance becoming increasingly and worryingly common, it could be that in a decade or two, antibiotics may be replaced by using bacterial adjustments to an individual's gut to improve their immunity to disease.

FIVE QUICK FACTS

 1 **YOU CAN'T TAKE A PULSE WITH YOUR THUMB**
The princeps pollicis artery runs into the thumb, giving the thumb its own strong pulse. That's why a doctor takes your pulse using the index and middle fingers—the thumb's pulse would interfere with the reading.

 2 **MOST PEOPLE BLINK BETWEEN 15 AND 20 TIMES PER MINUTE**
And their eyes are closed for an average of one-tenth of a second per blink. At the fastest rate, you could be blinking almost 20,000 times a day.

 3 **SHOES AREN'T GOOD FOR YOUR FEET**
In a 2007 study, it was found that the healthiest feet belonged to people who never wore shoes; the more regularly the subjects wore shoes, the more it threw out their natural gait.

4 **IT *IS* POSSIBLE TO DRINK TOO MUCH WATER**
It's good to stay well hydrated, but if you drink too much, too fast, your kidneys can't process your intake: your extremities will swell up and you'll feel sick. Everything in moderation!

 5 **INUIT PEOPLE HISTORICALLY GOT VITAMIN C FROM NARWHALS**
In an environment where the essential vitamin is hard to find, one square inch of narwhal skin contains as much vitamin C as an ounce of fresh orange.

67 WHERE DO DEAD BODY CELLS GO?

We're accustomed to hearing (and being revolted by) the fact that the dust that accumulates around our homes is largely made up of human skin cells that have been sloughed off in the course of day-to-day activity. But what about the cells inside the body? How long do they live, and where do they go when they die?

HOW CELLS DIE

Cells in your body die at a high rate. Scientists estimate around 50 billion perish in a single day. They can die in two ways— by apoptosis or necrosis. Most types of cells have a fairly set life span, and apoptosis is what might be called a cell's "natural" death: When it has outlived its usefulness, it begins to dismantle itself from within.

Proteins inside the cells called caspases prompt the production of enzymes that destroy the DNA of the cell and start to break it down. As part of the process, the cell becomes "leaky"; it sends out messages to the specialist cleaner cells, called phagocytes, which will deal with the debris of the dead cell.

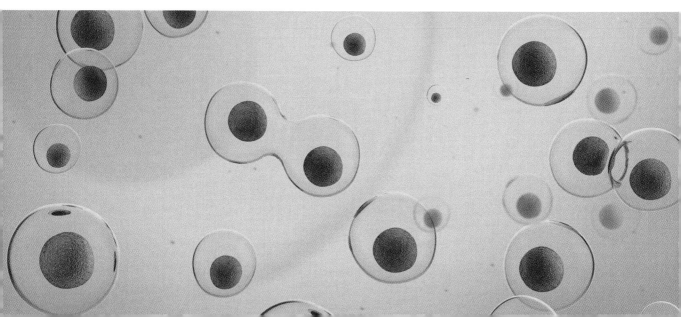

WHEN GOOD CELLS GO BAD

Sometimes the DNA of cells isn't disposed of properly—this can happen when necrotic cells die, and also sometimes if the phagocytes have had too much work to do. If this happens, it may provoke the autoimmune system of the body to overreact, in some cases causing serious conditions such as lupus, anemia, and arthritis.

Death by necrosis is less well regulated. It happens because a cell has been damaged in the course of a trauma to the body, such as an external injury or an infection. The sudden death isn't necessarily neat and self-contained like apoptosis. The dying cell, which bursts rather than leaks, won't send out the same signals as an apoptotic cell,

so it may not be so easy for the phagocytes to detect and remove it efficiently after it dies—although they will still "collect" a necrotic cell for breakdown. The chemicals released by necrotic cells can also trigger areas of inflammation in the body.

THE BODY'S TRASH COLLECTORS

The phagocytes that help to clean up after dead cells are two kinds of specialist white blood cells: neutrophils and macrophages. Although they're often referred to as being the trash collectors of the human body, their role is really more akin to that of very efficient recyclers: They literally engulf the remains of a dead cell and help to degrade them so that their component parts can be reused by the body. They're produced in your bone marrow, then travel in your blood to where they're needed. When a dead or dying cell is carried into their orbit, they'll move swiftly to gobble it up.

68 COULD YOU LIVE ON A VAMPIRE DIET?

Dracula and a whole clan of Cullens did it—although of course they're fictional. If you wanted to, though, would it be possible to live on an all-blood diet?

ANY OLD IRON, AND OTHER CONCERNS

The difficulty of working out the pros and cons of vampirism is that they're highly hypothetical. In the absence of genuine vampires, any facts and figures have to be speculative. The potential pros (you wouldn't want for protein) are wiped out by some pretty hefty cons (if you discount iron, blood contains negligible amounts of many vitamins and minerals). But if you were hell-bent on going gothic in this particular way, here are some of the other things that you'd be up against. You'd need to put away quite a lot to meet your energy needs. It's estimated that human blood contains 430–450 calories per pint (the amount of a typical blood donation). So an adult male would need to drink about 6 pints per day to get his 2,500 calories. Women could manage on slightly less.

Blood is very high in iron, but your system—which can tolerate up to 45 milligrams daily—could probably cope (although if you did overdose, it could cause a life-threatening condition called hemochromatosis, which leads

BETTER FOR BATS

Just three mammal species are known to live on blood alone—and all three are different kinds of vampire bats (as well as the common variety, there's also a hairy-legged vampire bat and a white-winged one). Unlike humans, vampire bats have evolved to cope with this specific diet. They boast supersharp teeth to cut through skin, anticoagulant saliva—thanks to the presence of a glycoprotein with the wonderful name of draculin—to allow their victim's blood to flow freely enough for them to lap it up, and an exceptionally unusual gut microbiome, which, among other things, features around 280 bacteria that would cause serious illness in other animals.

to a whole raft of problems, including heart and liver failure). Salt would be a concern, though—blood is very salty, with around 0.33 ounces in every 2 pints, which would add up to almost 1 ounce a day—way, way above the 0.25-ounce daily intake generally recommended. And you'd need to take a supplement to reach your recommended daily allowance of vitamin C, too: Human blood contains under 566 milligrams in 2 pints, so your 6 pints would give you around 1,400 milligrams, which is well short of the recommended daily 42,500 milligrams. Even if you're undead, you wouldn't want to risk developing scurvy.

ON THE PLUS SIDE

With their routine of sleeping through the daylight hours, there's one aspect of vampires' lifestyle that is right up to date. Humans in the twenty-first century are often told that they're chronically short of sleep, and current thinking is that you need at least eight hours a night—possibly as many as ten. So the vampire average of at least ten hours a day (naturally they'll be awake through the night) would be one healthy habit you could emulate.

69 WHY CAN'T YOU TICKLE YOURSELF?

Small children can be driven into ecstasies of hysterical pleasure by being tickled; as we age, we tend to stop finding the sensation so enjoyable (maybe we become more fearful about losing control or dignity?)—and some of us actively hate it. However, one thing every child knows is that it's nearly impossible to tickle yourself. Why?

A QUESTION OF EXPECTATION

Although it seems like a simple question, this raises some quite complicated issues about how you experience yourself as a separate entity from others. The idea that you know that you're you is so entrenched that it appears obvious to us, but it's impossible, so far, to give any form of artificial intelligence the same sense of self, so it's clearly more complex than it might appear. An experiment conducted at the Institute of Cognitive

Neuroscience at University College London looked at the different brain responses elicited when people were tickled by others, and when they tried to tickle themselves.

WHO'S TOUCHING YOU?

The experiment found that the brain creates a very clear map of your movements that simultaneously sends signals to the somatosensory cortex, where touch sensations are experienced, to "warn" it that the movements are your own and not anybody else's, and to the anterior cingulate cortex, which processes pleasurable sensations. The somatosensory cortex message ensures that you won't react when your own hand brushes your knee in the way that you would if it were someone else's hand. You know where your own touch will land and you have an expectation of it, whereas you don't have the same expectation of anyone else's movements.

The same study tried to set up a situation in which its subjects could fool their internal system and learn to tickle themselves. One experiment required them to move a lever that would cause their hand to be stroked—but at minutely different intervals. It proved that they could tickle themselves once an unpredictable delay was introduced, because they could no longer precisely anticipate when the ticklish trigger was going to arrive.

TWO KINDS OF LAUGHTER

Scientists have identified two categories of tickling—knismesis and gargalesis. Knismesis is the light "ticklish" sensation you get when a feather brushes against your hand—it's unlikely to make you laugh. Gargalesis, on the other hand, is what might be called "heavy" tickling: The kind that drives you into fits of semi-reluctant laughter. Not only are there two types of tickling; there are also two kinds of laughter. When heavy-handed tickling does make you laugh, it won't be in the same way you might laugh if you heard a funny anecdote. A study at the University of Tübingen, Germany, in 2013 found that while both kinds of laughter are felt in the Rolandic operculum (an area of your brain that controls emotional responses and the way your face moves in reaction to them), only the tickling-induced laughter also prompts a response from the hypothalamus, the region that can trigger your adrenaline-loaded flight-or-fight response and alert you to danger.

70 WHY DO OLDER PEOPLE SNORE MORE?

Maybe someone has never snored, but now that they're getting older, they've started. Or they were a moderate snorer, but in old age the noise has become both irregular and deafening. But what causes it, why does it get worse with age, and why is it so difficult to fix?

HOW SNORING STARTS

Snoring is a very common problem—a study in 1993 found that, of a large mixed sample of subjects, 44 percent of the men and 23 percent of the women tested were regular snorers. It is the result of muscles slackening while you're asleep. The muscles that line the airway and keep it open while you're awake relax as you drop off, narrowing or even partially obstructing the tube and making the airflow irregular. The air that usually flows smoothly and quietly down your throat and in and out of your lungs is disrupted by the now-variable width of the passage it passes down, and this creates irregular puffs

and blasts of air, which—when they hit the slackened muscles in the throat—result in the uneven-bellows effect of snoring. Heavier people will often snore more loudly than slimmer ones, too, because fat deposits stored in the walls of the airway vibrate in a similar way to loose muscles. And snoring may be aggravated as you age, as muscles will get floppier the older you get.

HOW TO STOP IT

There are numerous suggestions to help stop snoring, some more eccentric than others. Sleeping on your side, rather than your back, is less likely to obstruct the airway; one uncomfortable-sounding suggestion is to duct-tape a tennis ball to your back so that you don't turn onto your back while you're asleep.

FATAL SNORING

Snoring is irritating, but it can be dangerous, too. Bad snoring may be a symptom of sleep apnea, a condition that can cause sufferers to stop breathing altogether as they sleep. This may happen very frequently—in really bad cases, every few minutes—and as the body becomes aware that it's running short of oxygen, it startles the sleeper awake. This can cause disruption, sometimes disastrous disruption, to sleep patterns. It's even been implicated in a higher incidence of heart attacks and strokes, so if diagnosed it should be taken seriously.

DIDGERIDOO THERAPY

What's the strangest solution to snoring ever proposed? Learning to play the didgeridoo can help. A study carried out in 2006 in Switzerland gave patients at risk of sleep apnea regular didgeridoo lessons across a couple of months. Subjects were required to practice for at least twenty minutes a day, five times a week. And the results were positive—not only was daytime drowsiness (a common symptom of sleep apnea) reduced, but subjects' partners reported greatly decreased disturbance overnight, too. Why should it work? To play the didgeridoo, the musician must be able to practice circular breathing—that is, to inhale through the nose while controlling airflow into the instrument from the mouth—and this tightens and tones the muscles of the airways, making it less likely that they'll "flop" overnight.

71 DOES YOUR WHOLE BODY DIE AT THE SAME TIME?

It may be comforting to think of death as instant, but it actually takes a little time to change status from living to, well, not. After your heart stops beating, oxygenated blood stops circulating and the cells most dependent on it die first.

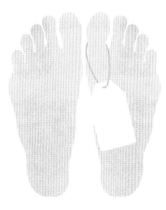

INSIDE OUT

The cells of your internal organs go fast (that's why it's crucial to harvest kidneys or a liver within at most 30 minutes of death), while your skin cells, at the other extreme, can live for much longer. If you were a skin donor rather than a kidney donor, doctors would have up to twelve hours to take the donation.

THE LIGHT AT THE END OF THE TUNNEL

What about the bright light so many people report moving toward at the moment of death (that's the ones who "came back")? There may be a scientific explanation. Your brain is one of the last parts of you to shut down, and when it is starved of oxygen, one side effect before you lose consciousness is tunnel vision. Combined with a sudden closedown as you black out, the effect might be very like moving down a tunnel toward the light.

WHY DOES DEATH SMELL SO BAD?

The horrible scent of putrefaction derives from a cocktail of chemicals, most of them by-products of bacteria that move in to colonize a cadaver. Two of the smelliest culprits are the evocatively named putrescine and cadaverine, both naturally produced by the breakdown of amino acids in the body after death. One lab worker suggested that a quick squirt of the two combined and you'd smell "dead" enough to ensure your survival in a zombie apocalypse.

QUIZ
THE HUMAN BODY

Think you know everything about your own body? Let's see if you can tickle your brain cells and remember what you have learned in this chapter.

QUESTIONS:

1. In a baby, where would you find the fontanelle?

2. How tall was Napoleon?

3. How much bacteria are you carrying in your gut?

4. Where would you find the Rolandic operculum?

5. Which musical instrument might help you to control your snoring?

6. What makes putrefaction smell so bad?

7. What tidies up the detritus from dead cells internally?

8. Do humans grow consistently taller over time?

9. How many human body cells do you lose in day?

10. Is there more than one way to tickle someone?

Turn to page 247 for the answers.

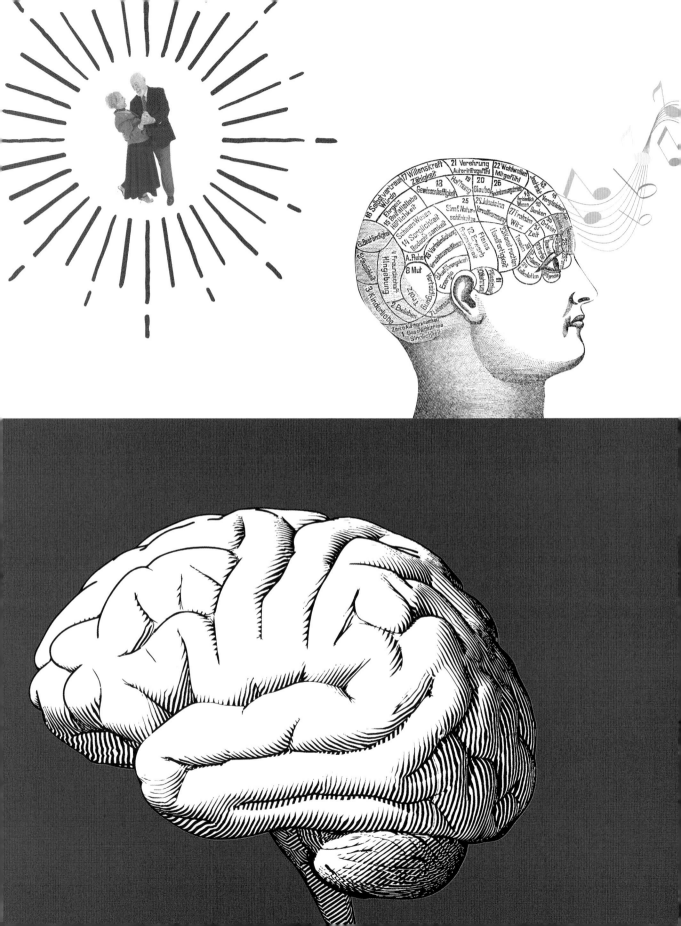

THE HUMAN BRAIN

72 COULD A HUMAN BRAIN BECOME FULL?

It's a familiar sensation: Your brain feels stuffed full of information, and you can't quite find the specific piece of information you want, however hard you try. But could your brain really fill up . . . to the point where it wouldn't be able to take in any more?

A LIBRARY NEEDS A FILING SYSTEM

The short answer is no—because the brain isn't a gas tank. It operates in a far more complex way than a simple "fill it up" mechanism, and it's been known for a long time that memory, key to ensuring that we think and act appropriately in different situations, is one of the most complicated systems of all, involving a number of different areas of the brain. Huge amounts of information aren't much use without an accurate method of calling up specifics (after all, the largest library in the world wouldn't be very effective without an efficient filing system).

In the human brain, one of the functions of the hippocampus, the horseshoe-shaped structure deep in the brain, is storage. But when you need a specific memory, the prefrontal cortex, which covers part of the frontal lobe of the brain, helps to filter what you need from what you don't. This means that you don't consciously have to scroll through memories every time you want a specific one.

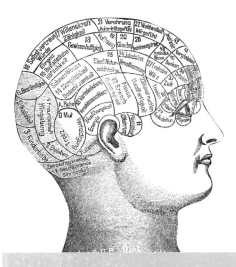

I'VE FORGOTTEN TO REMEMBER TO FORGET

Back in 1955 when Elvis was warbling the popular classic, no one thought that he was expressing a scientific principle rather than a romantic lyric. But it turns out that forgetting is a key part of keeping your memory efficient. With a constant "feed" of new material pouring into the brain, it's important that some of the earlier, lesser-used stuff can be put on a mental back burner to keep the memory agile.

FINDING THE "RIGHT MEMORY"

An article in *Nature Neuroscience* in 2015 recorded an experiment that seemed to show that the prefrontal cortex strained out less-relevant memories if the brain did a general search for something specific. If there were a number of similar memories, it appeared that the prefrontal cortex would ensure the right one was pushed to the fore. After the "right" memory had been found, if it was looked for again later, it attracted more brain activity than a less-used memory. Well-thumbed memories, taken out and "used" regularly, attracted the most brain activity.

SUPERMEMORIES

You might wish your memory was better—but what if you could remember every day of your life so far, in full, excruciating detail? People with hyperthymestic syndrome can do just that: Ask them what they had for dinner on a random date five years ago and they'll be able to tell you, accurately and without hesitation. Alexander Luria, a Soviet psychologist and one of the founders of neuropsychology, was one of the first to record the condition when he noted that "Mr. S," a subject he studied in Moscow in the 1920s and 1930s, was desperate to discard items from his all-too-inclusive memory, first writing them down on scraps of paper in the hope of getting rid of them, and then, when that failed, burning the paper to ashes (it didn't work). Now, it seems likely that Mr. S suffered from hyperthymesis, although the term wasn't coined until early in the twenty-first century.

73 WHAT ARE DREAMS FOR?

It's hard to say definitively what dreams are for. They've been the subject of both amateur and professional interest for centuries, yet despite being so widely studied, they remain obstinately open to a vast number of interpretations.

EVERYONE DREAMS

Even when you don't remember them on waking (it's believed that the dreamer won't have any recollection of 90 percent of their lifetime's dreams), it's estimated that you probably have somewhere between three and six episodes of dreaming every night. How do we know? Because many studies have established that between 20 and 25 percent of your sleeping time is spent in REM (rapid eye movement) sleep, and REM sleep indicates that you're dreaming.

THEORIES OF DREAMING

Sigmund Freud, the father of psychoanalysis, was the first modern great to put forward a theory of dreams. In *The Interpretation of Dreams*, published in 1899, he held that they were fundamentally expressions of wish fulfillment: everyone's chance to explore the thoughts that they might not dare express, or even consciously feel, when awake.

THE KNACK OF NIGHTMARE AVOIDANCE

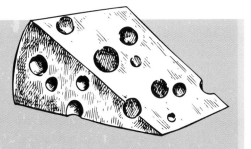

The old myth that eating cheese before bedtime will give you bad nightmares is just that—a myth. But there's some research that seems to suggest that going to sleep early may reduce the likelihood of bad dreams. In 2010, two studies, one from Turkey and the other from Canada, found that people who went to bed very late reported a higher incidence of nightmares. It's not clear why this should be, although one theory blames a natural rise in your cortisol levels in the early morning. On an ordinary schedule, this would be just before you wake up, but if you were very late to bed, the rise may happen while you're still in REM sleep—and could be a prompt for unusually vivid, bizarre, or just plain horrible dreams.

In the 120-odd years since his magnum opus was published, Freud's theories have been debated, dismissed, and reinstated in a recurring cycle. But even though some people disagree with him, and many have over the years, he set the scene for debate.

TRIED FAVORITES

There are literally dozens of extant theories on the purpose of dreams. Just three current favorites:

There's the snappily named activation-synthesis hypothesis. This holds that dreams don't really have an inherent meaning. Instead, they're a random selection from our thoughts and feelings that are constructed by electrical impulses in the brain. When we wake up, proponents of this theory hold, our conscious minds try to turn this jumble into "stories" in an attempt to make sense of them.

Another theory is that dreams are a by-product of the brain processing information. The idea goes that as we sleep, the brain works to make sense of, and store, all the information that it has acquired during the preceding day, and dreams are either the waste material from this processing or possibly even an as-yet-little-understood stage in it. There's also the threat-simulation theory. This explains dreams as a sort of simulation process. It suggests that threatening dreams give humans a mental rehearsal for difficult situations that might arise in real life—helping to ensure that they make the most pragmatic decisions when such situations arise when they're awake. This explanation handily covers the fact that animals other than humans dream, too.

74 WHY DOESN'T YOUR INTERNAL TEMPERATURE MAKE YOU FEEL HOTTER?

People differ in their responses to warm weather, but most would agree that when the thermometer hits 98.6°F on a tropically sunny day, they feel hot—even uncomfortably hot. But why, when that's the exact internal temperature of the human body?

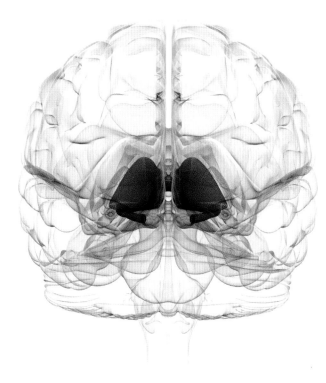

HEATING UP

Whether you're resting up or moving around, your body is in a constant state of activity, from your beating heart to your firing brain synapses. Even if you were the laziest couch potato in the world, numerous processes would still be carrying on automatically, and every metabolic process creates heat. So it's important for your body to be able to cool down efficiently.

A BUILT-IN THERMOSTAT

The hypothalamus, a small area at the base of your brain, is charged with regulating your body temperature. It helps to even up the heat created in the different areas of the body so that you're a more-or-less even temperature all over most of the time. But when it comes to getting rid of excess heat into your outside surroundings, you are still dependent, to an extent, on the temperature of your habitat. When you're cold, it's the hypothalamus that sends out the instruction for you to start shivering, putting the body into motion to warm you up; when you're hot, on the other hand, it will direct your body to sweat and to increase the amount of blood flowing through the vessels closest to the skin's surface, to give your body the best chance of cooling down.

IT'S NOT THE HEAT, IT'S THE HUMIDITY

It's the catchphrase that gets repeated over and over again in muggy, sticky weather. But why do humidity levels make such an unwelcome difference to the way you experience the heat? It's because humidity affects your ability to sweat effectively. Usually, when you break a sweat in the heat, you cool down as the liquid evaporates off your skin. But if you're sweating in air that is already saturated with moisture, the sweat won't evaporate, and you're left feeling hot, damp, and uncomfortable.

BABY, IT'S WARM INSIDE

It's a rule of physics that when two objects of different temperatures are brought together, the one with the higher temperature will transfer heat to the one with the lower temperature. This is why a cup of hot coffee left out on the kitchen counter will gradually cool to room temperature. In the same way, your body will leak heat out into cooler surroundings. If the external temperature becomes equal to, or even higher than, your internal one, the heat has nowhere to go, so you will find it increasingly difficult to cool down.

75 WHY CAN SOME PEOPLE HEAR COLORS?

It's called synesthesia, and is probably best described as an involuntary combination of the senses. Hearing colors is probably the best-known example, but it can take a number of forms: experiencing sounds as smells, matching textures with emotions, and sensing words as tastes are just three others.

BRAIN GAMES

It was first identified in the nineteenth century and it's been studied at points ever since, but synesthesia hasn't yet been explained, although there are several different theories for what causes it. There's a tendency for it to run in families, which implies that there may be some genetic link. One argument is that it results from neural connections between those areas of the brain dedicated to the senses that are usually distinct (sound having additional connections to sight, for example). Another idea is that everyone has the potential neural pathways to experience synesthesia, but that, for reasons as yet unknown, the brains of only a small percentage of people use them. However, it's clear that a lot more research will need to be done before there's a theory on which everyone can agree.

LANGUAGE OF THE SENSES

A recent study made an interesting connection between language and synesthesia. People who learn a second language early in life but who aren't bilingual (that is, they don't learn to speak two languages simultaneously from infancy) are more likely to be synesthetic. This suggests that synesthesia may be some kind of mental reaction to complex learning processes, such as mastering grammar or learning music—a deep-brain equivalent to those kindergarten alphabets where, say, all the Bs are red, while the Rs are all colored blue.

FIVE QUICK FACTS

 1 LENIN'S BRAIN WAS PROBABLY THE WORLD'S MOST STUDIED

After his death, Vladimir Ilyich Lenin's brain was cut into 30,000 wafer-thin slices and saved for posterity at the Moscow Brain Institute. Stalin's brain was kept just down the hall.

 2 IN MODERATE DOSES, COFFEE IMPROVES YOUR BRAINPOWER

Caffeine is a cognitive enhancer—boosting your alertness, sharpening your memory, and speeding up your reactions. Don't go over 400mg daily, though, or these benefits may be wiped out by jittery side effects.

 3 WITHOUT SALIVA, YOUR FOOD WOULD BE TASTELESS

Every mouthful of food has to be partially dissolved by your saliva before the receptors on your taste buds can register its flavor and pass the information to your brain.

 4 THE BRAIN HAS NO PAIN RECEPTORS

Nociceptors are our neurons' response to stimuli; they need the brain to interpret them for us and to "read" them if they register pain, but the brain itself doesn't contain them. That's why it's possible to perform brain operations on a patient while they're awake.

5 A BIGGER BRAIN DOES MAKE YOU MORE INTELLIGENT, BUT NOT BY MUCH

In a study run jointly by the Universities of Amsterdam and Pennsylvania in 2018, it was found that people with larger brains did perform better in cognitive performance tests, but on a variable of only about 2 percent.

76 CAN PUZZLES STOP ALZHEIMER'S?

It's one of those pieces of received information that's probably been quoted to you many times: Keep your mind agile, do plenty of crosswords and sudoku, and it'll reduce your chances of getting Alzheimer's disease in old age. But is it true?

ENVIRONMENT VS. HEREDITY

Only partly. This is an area that has seen several studies, and just as many inconclusive results. However, there appears to be a grain of truth in it. Around one in five people carry a gene variant—APoE4—which doubles their risk of developing Alzheimer's in later life. And for those people, doing crosswords, sudoku, and any other mentally stimulating activity has been shown to help delay the buildup of abnormal protein deposits (often referred to as "plaques") on the brain—and these have been strongly implicated in the development of Alzheimer's. Conclusion: Keep doing crosswords, sudoku, and any other activity you find stimulating. It can't do you any harm.

KEEP DANCING

There's no need to limit yourself to puzzles, either. A study of older people carried out by the Karolinska Institutet in Stockholm took a sample of more than 1,200 subjects aged between sixty and seventy-seven in Finland, and turned the lives of half of them upside down—in a positive way—across a two-year period between 2009 and 2011. They were given exercise and aerobics classes and brain training with graded computer games, while their diet was overhauled to include plenty of vegetables, fish, and healthy oils. Perhaps unsurprisingly, at the end of the period, the revamped sample did substantially better than the control subjects in a range of tests, including an 85 percent improvement in tests of their brains' organization of thought processes, and a staggering 120 percent in the speed at which they could process information. The control group, who had

been given straightforward health advice of the kind typically offered in a doctor's office but nothing else, remained at the cognition levels of their original tests.

HIGHER EDUCATION

There are increasing indications that a higher degree of education may also help the susceptible to dodge a bullet when it comes to Alzheimer's. Another study by the Karolinksa Institutet found a correlation between the odds for contracting Alzheimer's disease and the level of education reached by the subject. Each year of additional education past the primary stage resulted in lower odds for Alzheimer's, all the way up to completing a university degree.

EXERCISE CAN BE FUN

Asked what the takeaway from the study was, Professor Miia Kivipelto, who led the research, suggested that in the absence of "management" to ensure an all-around lifestyle change, older people should take on extra or different activities one at a time. Her top recommendation for a new activity? Dancing. It's often a less intimidating option for the elderly than going to the gym, plus it can offer fun and some social interaction, as well as exercise. So if you've already taken on some mentally stimulating options, consider adding a weekly dance class, too.

77 HOW MANY ORGANS COULD YOU LIVE WITHOUT?

We all know someone who's had their appendix out (and good riddance, they may feel—an inflamed appendix is famously painful, and it's an organ with no known function anyway), but how many other organs could you live without altogether?

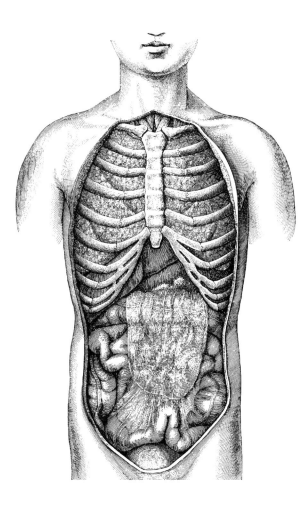

OPTIONAL ORGANS

You can lose one kidney without too many problems—provided that the other one is fully functional, it will do its best to keep up with the work, although you can help it by following a healthy diet and eschewing alcohol. The same goes for the lungs. Typically, a person uses only around 70 percent of their potential lung capacity with two lungs, so losing one doesn't mean losing half their full capacity, although the remaining lung will have to work harder to compensate.

HOW MUCH CAN YOU LOSE?

Of course you can also manage without a uterus (in a woman) or testicles (in a man), although you won't be able to have children. And it's possible, too, to cope without a complete digestive system— you could live without your stomach or much of your colon, though not without discomfort or inconvenience. Taking out your spleen would mean an increased susceptibility to infection because it's a part of your immune system, and if your gallbladder had to be removed, you'd no longer have the extra bile that it stores to help you digest fat, so you'd have to avoid excessively fatty meals.

You would be exceptionally unlucky to lose all, or even most, of these organs—but theoretically, at least, you would be able to survive.

HALF A BRAIN

You would probably be able to handle the suggestion that someone remove, say, your gallbladder with reasonable equanimity. But what if someone were to propose taking out a large chunk of your brain? Much less acceptable. Yet there's an operation called a hemispherectomy that removes one hemisphere of the brain, and patients can live quite well for years with just the remaining half. The first hemispherectomy was performed at Johns Hopkins University in 1928, and

the operation has since become a specialty there. It's a last-resort operation, usually carried out on patients who suffer so badly from seizures that daily life has become impossible, and the huge majority of patients are young children.

YOU CAN'T TAKE MORE THAN HALF

John Freeman, an eminent professor of neurology at Johns Hopkins, quipped, "You can't take more than half. If you take the whole thing, you've got a problem." Even the operation to remove half is not without side effects, although they're not so bad as one might suppose: Patients usually lose motor function on the opposite side of the body from where the brain was removed, and may have speech problems, but can generally operate normally in most other ways. What happens to the skull space left by the vacated brain? It doesn't stay empty; the obliging body quickly fills it with fluid.

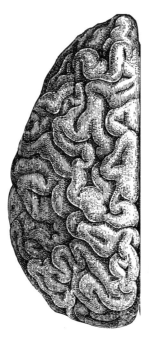

78 WHAT WAS THE GLASS DELUSION?

If you've ever even thought about it, you might assume that madness doesn't change much with time. But there's evidence that some mental disorders belong to their era—after all, people couldn't fantasize about being radioactive before anyone knew that radiation existed. Perhaps the most extraordinary example of this is the glass delusion.

A SHATTERING CONDITION

Those who suffered from the condition believed they were made of glass. This gave them terrible problems. One of the first cases to be documented, at the beginning of the fifteenth century, was King Charles VI of France. Handsome, magnetic, and initially a popular ruler, for the last thirty years of his reign he had lengthy periods when he believed his body was made of glass. He spent most of his time in bed, cushioned with thick layers of linen and wool; when he had to move, he wore a kind of corseted robe sewn with iron stays, to ensure that he wouldn't "break." Nor was he the only aristocrat to be thus afflicted. An even more convoluted case, and one of the last to be recorded, in the 1840s, was that of Princess Alexandra Amelie, daughter of Ludwig I of Bavaria, who maintained that in childhood she had swallowed a grand piano—but one made from glass. As a result, she moved gingerly through her parents' palaces, easing herself around corners, terrified that she would smash the glass piano into bits.

Although the records show that glass delusionists were well-born and highly educated, this may simply be because sufferers from less eminent backgrounds were ignored, laughed at, or locked away (and not included in the records). For about two centuries, the glass delusion was a widespread problem, afflicting hundreds of people, widely recognized and written about, before gradually fading away.

MADE FROM CORK

The English academic Robert Burton includes it in his book, *The Anatomy of Melancholy*, published in 1621. "[They believe] that they are all glass, and therefore will suffer no man to come near them . . ." he wrote (although he also includes people who believe they are made from cork, and others who think they have frogs living inside them, so it's part of a fairly comprehensive list). It was a well enough known condition, too, to appear in Miguel de Cervantes's 1613 story *El Licenciado Vidriera* (*The Glass Graduate*).

MADNESS MOVES ON

The glass delusion is almost unheard of today. The first cases were noted at the beginning of the fifteenth century—when clear glass was a novelty. Professor Edward Shorter, a scholar of the history of medicine at the University of Toronto, has pointed out that, before the heyday of the glass delusion, some people believed that they were made from pottery. After the glass delusion ceased to be reported, in the nineteenth century doctors wrote about obsessives who thought that they had turned into another newly minted material, concrete. Could it be that some forms of mental disorder simply land on a scientific novelty and adapt themselves? After all, today no one would be surprised to hear that someone suffering from a mental disorder believed that they were being influenced by hostile forces coming from their laptop.

79 IS STRESS EVER GOOD FOR YOU?

We're accustomed to being told how stressed modern life renders us—which comes with the assumption that stress is necessarily a bad thing. But are there any circumstances in which stress might be good for you?

GOOD STRESS, BAD STRESS

Contrary to what lifestyle magazines might tell you, high levels of stress aren't new: Humans have been living stressful lives since the first *Homo sapiens* had to face down a Neanderthal as they competed for the last local mammoth. But the word itself is used as a catchall that can mean anything from a slightly over-busy work schedule or social calendar to dealing with serious illness or death. To distinguish between good and bad stress, scientists tend to refer to "eustress" (good)—the

kind of stress that provides the impetus to stretch yourself and make an effort—and chronic stress (bad), which acts as a long-term drag on your mood and health.

In humans, stress sets up an early warning that is detected by the hypothalamus at the base of the brain, which in turn prompts the production of a number of hormones, including adrenaline and cortisol. You've probably found that in a suddenly stressful situation you think more clearly and are more decisive, and this is because of the

call-to-action effect of freshly released hormones. Adrenaline (also called epinephrine) is raising your heart rate and blood pressure, while cortisol is prompting the release of more glucose into your bloodstream and simultaneously slowing down body functions that don't need your attention in an emergency. Not only that, but stress can send your immune system into high alert, ready to defend you if need be. A slight degree of stress will motivate you, encourage you to focus on your immediate situation, and energize you— all things that also tend to make you feel upbeat and positive.

TOO MUCH OF A GOOD THING

Over a long period, though, if stress becomes chronic, it loses its positive qualities. Almost all the factors that make a little stress a good thing are turned on their heads with prolonged stress: It can mess with your immune system, making you prone to infection, affect the efficiency of your thinking, and, in extreme cases, become a contributing factor to depression.

THE HERE AND NOW

Living in the present isn't just a mindful exercise. A study undertaken by Harvard University, published in Science magazine in 2010, looked at the degree to which our minds wander. It gathered results by using a phone app that telephoned a wide selection of people (5,000, located in eighty-three different countries) at random moments during the day and asked them three questions: "How are you feeling?" "What are you doing?" and "What are you thinking?" The results showed that people who were fully focused on what they were doing reported feeling happier than people who weren't, even when the latter were thinking about enjoyable things. We hear a lot about the way in which animals live "in the moment" (scientists believe this is so because they don't actually have the capacity to do anything else), but if the human animal lived in the moment, too, it's likely that they might be happier—and less subject to chronic stress.

80 WHY CAN AMPUTEES STILL FEEL THEIR MISSING LIMBS?

Ambroise Paré, a notable French surgeon, was the first to write about phantom limb syndrome in the mid-sixteenth century. It's the phenomenon by which amputees feel a nonexistent limb is still there, and is sometimes painful, too. So what's going on?

THE GHOST IN THE BRAIN

Phantom limbs occur in a very high percentage of amputees—some studies have estimated as high as 95 percent. The sensation that the missing limb is still present is so powerful that a subject may even move as though their leg or arm is there: If a ball is kicked toward them, for example, they'll move to kick it back, attempting to use the limb that isn't there. There are many theories about where

phantom limb syndrome originates, from the nerve endings near the amputation to messages within the spinal cord. The majority of scientists believe that it goes back to the brain and to the somatosensory cortex, which takes in and deals with information gained by touch. A number of studies have found that the brains of many amputees don't seem to have registered that a limb was missing: The area of the brain that would have

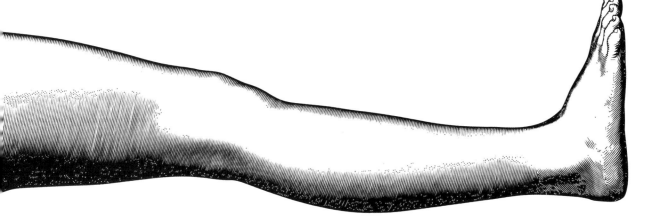

been responsible for moving and controlling it had often hardly altered at all. And it's possible that this "things as normal" situation in the brain causes the sensation of a phantom limb.

THE MIRROR TREATMENT

For a long time, the main treatment for phantom limb pain was conventional painkillers, which seemed to have very hit-and-miss results. But in the early 1990s, the neuroscientist V. S. Ramachandran, working at the University of California, San Diego, developed a new treatment to help amputees. It was a simple mirror, placed opposite an amputee's remaining limb, so that the reflection made it look as though the missing limb was still there. Amputees did a series of exercises, watching the mirror— and reported that "seeing" the missing limb exercising led to a decrease in pain. There's still a good deal of debate about how the mirror treatment works, but it's believed that the brain may be sending pain messages connected to the missing limb as the result of a sense that something is wrong, and that offering a visual perception that the limb is there may trick it into believing all is well.

NELSON'S ARM

In 1797 Lord Nelson lost his right arm during a naval assault on Tenerife. He was a tough patient; legend has it that he was back issuing commands within half an hour of the amputation, and a couple of weeks later his surgeon, James Farquhar, noted how cleanly the stump was healing. Yet Nelson told friends that he felt his arm, and the pain of it, as clearly as when it had been there. In addition, he sometimes felt fingers touching the palm of his hand—even though the hand wasn't there. Instead of treating the experience as a problem, however, he took the "presence" of his arm as proof positive that the human soul existed: If his arm was enjoying an afterlife, he said, why should the rest of him not follow?

81 HOW MANY WATTS DOES YOUR BRAIN RUN ON?

Your brain uses more energy than any other body organ—anything up to 20 percent of the total amount your body generates. So what does that mean in wattage terms? Could your brain energy power a light bulb, for example?

SHARP BRAIN, DIM BULB

The answer is pleasingly precise: 12.6 watts. That figure was published in an article by Ferris Jabr in *Scientific American* in 2012. He made the calculation by converting the calories used by an average person's daily resting metabolic rate (around 1,300 kilocalories) into joules, then from joules to watts, before finally dividing the total (63 watts) by five (20 percent). That's not very much in commercial energy terms—think of the dim light supplied by, say, a 25-watt bulb. Yet that's double the wattage of the human brain.

WHAT'S ALL THAT ENERGY FOR?

A 2008 study at the University of Minnesota Medical School found that two-thirds of the brain's energy requirements are used in communication—ensuring neurons send the signals necessary to meet all the jobs the brain does; the remaining third is dedicated to maintenance, more of which takes place during the body's downtime.

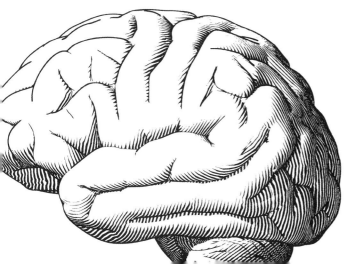

QUIZ
THE HUMAN BRAIN

Use your brilliant brain to recall everything you have learned in this chapter! You may remember more than you think.

QUESTIONS:

1. How much of your brain could you live without?

2. Which organ regulates body temperature?

3. What is hyperthymestic syndrome?

4. How much of your brain's capacity is used for communication?

5. Princess Alexandra Amelie of Bavaria suffered from a very strange delusion. What was it?

6. What would you be unable to do if your gallbladder were removed?

7. What part of the brain produces the hormones adrenaline and cortisol?

8. Why does hot and humid weather make you feel warmer than hot weather?

9. What is eustress and why is it good for you?

10. Does synesthesia always mean that you can hear colors?

Turn to page 248 for the answers.

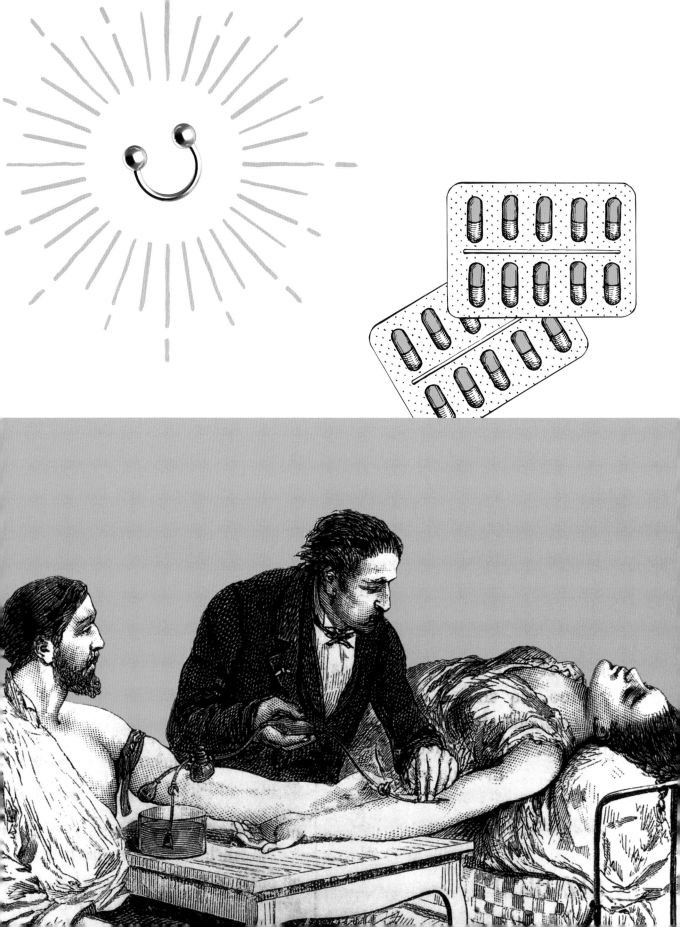

MEDICINE

82 HOW COULD ANTIBIOTICS STOP WORKING?

Today it's hard to imagine life before antibiotics, even though the widespread use of penicillin goes back only to the mid-1940s. Their discovery transformed medicine, so it's frightening to be told, around seventy-five years after they were first introduced, that they're ceasing to be effective.

NATURAL RESISTANCE

Antibiotics act against bacteria either by killing them outright or by preventing them from reproducing. And although it may seem to have developed very quickly since humans discovered how to make use of them, bacterial resistance to antibiotics is a natural occurrence. We're partly to blame ourselves for having used vast quantities of antibiotics in every imaginable situation: We've given the bacteria plenty of opportunity to practice.

KNOW THY ENEMY

Resistance happens because those bacteria least susceptible to the effects of antibiotics are the ones that survive and reproduce. It's survival of the fittest at a bacterial level. Over time, as only the toughest bacteria survive, they develop strains that respond less and less to antibiotics, until eventually they are fully resistant, and the antibiotics won't work. And some have developed multiple resistances to a range of antibiotics, creating "superbugs" that are very hard to treat. Some bacteria can actually destroy the antibiotic—they produce an enzyme called beta-lactamase, which breaks down penicillin, rendering it useless.

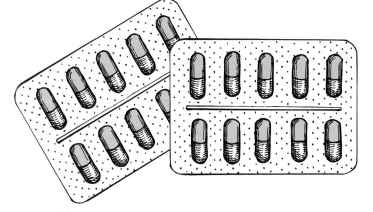

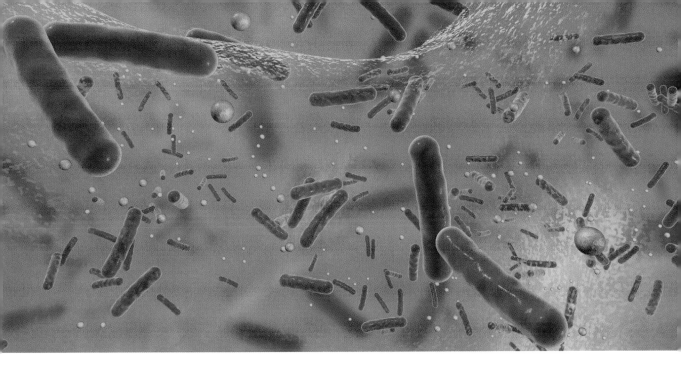

LOOKING INTO THE FUTURE

Some authorities believe that we have only a very limited time before the majority of antibiotics become useless against the majority of bacteria (some have estimated just a decade or so). This would take medicine back to the point at which ordinary infections became killers, and it would be difficult, if not impossible, to perform operations safely. Apart from stringently limiting prescriptions of those antibiotics that bacteria find it hardest to fight against, to maintain their effectiveness for as long as possible, what can research do to protect us? Researchers are looking at other ways in which our bodies might help us to protect ourselves. The immune system is the focus of a lot of work, in the hope that there might be a way to teach it to protect us better against bacteria in the first place, making antibiotics redundant.

COULD THE PROCESS GO INTO REVERSE?

It's possible that the resistance to antibiotics could gradually disappear, but only if bacteria didn't come into contact with them over a very long period. If the threat that antibiotics pose to bacteria were to vanish altogether, scientists believe that bacterial defense tactics would gradually also erode, but it could take a very long time without any antibiotic use at all—at least decades, and some say even centuries—before they would become effective again to any useful degree.

83 WHEN WAS THE FIRST OPERATION?

The first operations for which there is archaeological evidence were trepanations, and they date back as far as late Paleolithic times—which means some may have been carried out around 12,000 years ago, and across a whole range of cultures in Africa, the Americas, Asia, and Europe.

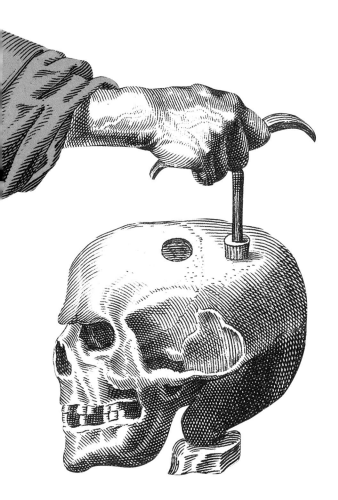

HOLES IN THE HEAD

When you consider that a trepanation (or, as today's surgeons would call it, a craniotomy) involves literally drilling, or, worse, scraping a hole in your head, and that the only available tools in those early days would have been flints, or the sharp edges of seashells, the idea of undergoing one sounds less than appealing. So how and why were they carried out?

The first person to take an interest in an ancient skull with signs of trepanning was an American diplomat and archaeologist named Ephraim (E.G.) Squier, who, in the 1860s in the course of a visit to Peru, was shown an ancient Inca skull bored with an evidently deliberate hole. He brought it home with him and it became the subject of intense debate: Had the subject survived the operation (had it, in fact, been performed pre-mortem at all)? Squier noted that the hole in the skull showed signs of having partially healed over. He concluded that it had been done on a living patient, and, moreover, that the patient had survived it.

REPAIR OR RITUAL?

The question "Why?" remained, and the debate broadened as, throughout the nineteenth and early twentieth centuries, more drilled skulls, many very ancient, were found all over the world. At the same time, scholars were pointing to the written works of the surgeons of the ancient world, notably Hippocrates and Galen, both of whom advocated trepanning—originally named for the "trepan," or drill, which was used for the operation in ancient Greece and Rome—for some brain conditions and injuries. Even using a drill rather than a flint or a shell, it sounds tricky. Early writers urged that a bowl of water be ready to plunge the drill into when it became too hot, and commented that it was hard work to bore all the way through the skull without inadvertently going into the brain itself. From these sources, it might have been deduced that all trepanning was carried out for medical reasons. But while some skulls showed evidence of wounds or trauma, others seemed to have been carried out on skulls that showed no sign of previous damage, raising the question: Did trepanning have ritual purposes, too?

LETTING IN THE LIGHT

Trepanning has always involved letting something in or out of your head. When it wasn't a straightforward procedure to relieve pressure on the brain, it seems often to have been a case of letting devils out or enlightenment in. Aficionados—perhaps surprisingly, there are some contemporary enthusiasts—believe that it restores the natural lightness and well-being that you feel in childhood, before the bones of the skull fuse and harden. In 1970, Amanda Feilding, a British artist and drug policy reformer, performed the operation on herself while filming it. Afterward, she said, she wrapped her head in a scarf, ate a steak to replace the blood she'd lost, and went out to a party. She claimed that the mental benefits were subtle, but positive. In interviews she stuck with official medical advice: Don't try this at home.

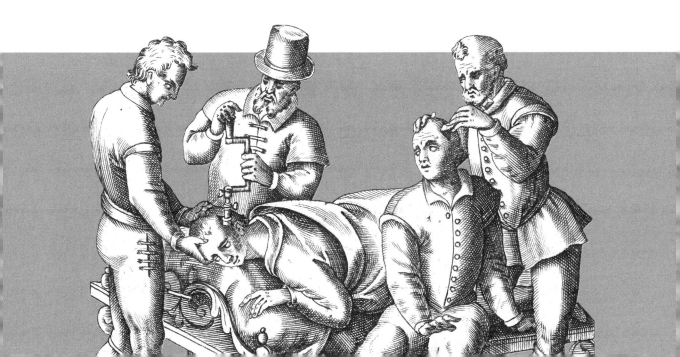

84 HOW DO DISEASES DIE OUT?

Generally, diseases have proved able to resist human efforts to eradicate them—influenza, malaria, and measles, after all, are still with us. The exception is smallpox, which was a scourge all over the world, killing millions until it was finally wiped out in the 1970s.

EARLY BEGINNINGS

Smallpox is a very ancient disease. When unwrapped, the mummy of Ramses V, dating back to the twelfth century BCE, was found to be marked with the scars of smallpox blisters. Of those who contracted it, 30 percent died, and even those who survived were often left blind and almost always with a disfiguringly scarred and pocked face.

The human fight against smallpox began with a process called variolation, of which the intrepid world traveler Lady Mary Wortley Montagu was an early advocate. Introduced to Europe in the eighteenth century by Turkish doctors, variolation was kill-or-cure risky: Pus was taken from smallpox blisters and someone who hadn't had smallpox either breathed it in or had

it placed into a cut on their arm. Most then had an attack of smallpox, but fewer died than if they had caught it naturally. In 1796, Edward Jenner developed vaccination, a way of guarding against a disease by deliberately infecting subjects with a similar but milder disease. He had noticed that milkmaids, who had invariably had cowpox, never contracted smallpox, so began deliberately "vaccinating" (injecting subjects with a vaccine made from cowpox lesions). Once there was a vaccine, humans had an effective weapon against smallpox.

THE DEATH OF SMALLPOX

Long after it had been eradicated in the richer countries of the world, smallpox raged on in the poorer ones. What finished it in the end was a global program of vaccination, accompanied by rigorous isolation policies around any sufferers. The World Health Organization announced a program of eradication in 1959, followed by one of "intensified eradication" in 1967. Eventually smallpox survived only in Africa, where a stringent vaccination campaign eventually saw it clinging on in just three countries: Somalia, Ethiopia, and Kenya.

COULD SMALLPOX RETURN?

Smallpox may have left the world stage, but it's not yet extinct—samples survive in a couple of laboratories, one in the United States in the Centers for Disease Control and Prevention in Atlanta, and the other in the State Research Center for Virology and Biotechnology at Koltsovo, Novosibirsk, in Russia. Originally, the idea was to keep samples for research purposes in the event of breakouts from unforeseen sources; today, they've long been the subject of international wrangling. There's the fear of biochemical warfare, of course: In the wrong hands, smallpox would be the ultimate chemical weapon. And stores of the vaccine no longer exist. In 1990, the World Health Organization allegedly destroyed 9.5 million of the last 10 million doses on the grounds that they were no longer needed.

The last natural case of smallpox occurred in Ethiopia in 1976, although there were to be a few further cases due to accident. Its eradication was announced by the WHO in 1980, in a triumphantly worded resolution: "The world and all its peoples have won freedom from smallpox . . ." it said, "[demonstrating] how nations working together in a common cause may further human progress."

85 WHICH INFECTIOUS DISEASE HAS KILLED THE MOST PEOPLE IN HISTORY?

The history of infectious diseases offers a grim roll call of mortality, and inevitably your mind goes to the last horrific outbreaks you heard of: Ebola virus, for example. But actually, Ebola ranks very low in what gallows humor might call the Grim Reaper awards.

HOW MUCH DO WE KNOW?

Obviously it's impossible to say with absolute certainty which disease has been the deadliest across the whole of history, because we simply don't have the necessary records. Historical statisticians enjoy informed speculation, though, and their labors have turned up some interesting facts.

Take the Black Death, for instance—the wave of bubonic plague that laid waste to large areas of Europe, Africa, and Asia for seven long years in the mid-fourteenth century. Some historical statisticians reckon it caused a total of 200 million deaths, while others go down as low as a still-shocking 75 million. In Europe alone, it's estimated that around 50 million people, more than 60 percent of the contemporary population, died.

TUBERCULOSIS VS. MALARIA

Other major killers through history include influenza, cholera, the now-eradicated smallpox (which claimed its last victim in a laboratory accident in 1978), and malaria. But which killed the most? Over the last two centuries, a period for which it's easier to make accurate estimates than the more remote past, tuberculosis has caused an estimated 1 billion fatalities.

But malaria is still likely to be considerably out in front—for a while it was believed that it had killed half the people who had ever lived on Earth (with an estimate of 50 billion deaths). We don't have accurate enough information to tell us exactly how large the world's population was at different points in history to back up that claim—but experts believe the estimate could be close.

NEW KIDS ON THE BLOCK

The older diseases—and mosquitoes containing malarial parasites have been found dating back around 30 million years—have spent a very long time claiming victims. At the other end of the scale, HIV and Ebola are both young diseases; neither has yet reached its fiftieth birthday. HIV is estimated to have killed around 36 million people since its emergence in the Democratic Republic of Congo in 1976. Ebola, which was "discovered" in the same year, has, despite its horrifying reputation, killed a comparatively tiny number of people overall: up to August 2018, fewer than 15,000.

EPIDEMICS—AND PANDEMICS

An outbreak becomes an epidemic when a disease spreads to several areas from a common source, in much greater numbers than would usually be expected. A pandemic is an epidemic on a global scale: it happens when an epidemic travels, first across countries, and ultimately across continents.

When COVID-19 was first heard of, it was at the outbreak stage, but it became a pandemic very quickly. It was the first in more than a century, since the Spanish flu outbreak in 1918. In September 2021, still mid-pandemic, the COVID-19 statistics stood at over 4.5 million deaths; by comparison, the Spanish flu killed somewhere between 25 and 100 million people. Epidemiological record-keeping has improved, so COVID-19 stats, while grim—and set to rise—are much less approximate.

86 WILL MEDICINE EVER BE TAILORED TO OUR DNA?

DNA is famously known as the "blueprint" that all living things have, but with an ever-increasing knowledge of the detail of how it works, what are the chances of doctors creating medicines that are a specific fit to our individual DNA profiles?

A NEW SCIENCE

To some extent, this is already possible. The comparatively new study of pharmacogenomics looks at ways to personalize medicine to maximize its effectiveness (and minimize drawbacks, such as bad reactions), using individual DNA. But although it is anticipated that within a decade or two, DNA profiling may become the norm in doctors' offices, personalized medicine is still at a relatively early stage.

WORKING WITH THE INDIVIDUAL

In genetic terms, most of our drug treatments are still surprisingly crude. When you are told you have a specific illness, you're usually given drugs that would be given to most other people with the same condition. If the first drug doesn't work, your doctor will try a second one. Our genetic profiles will give doctors the knowledge, in advance, of which treatments are likely to work and which won't "gel" with our genes.

Cancer is at the forefront of personalized medicine research because so many cancers are the results of genetic mutations. In 2011, the *Wall Street Journal* reported the number of cancer tumors that were specifically the result of genetic mutations that could already be targeted by tailored treatments. Melanomas got the most encouraging results with 73 percent of genetic tumors treatable, but even hard-to-treat lung and pancreatic cancers came in at 41 percent.

FIVE QUICK FACTS

 TRUE SILENCE CAN CREATE AUDITORY HALLUCINATIONS

In situations of unnatural quiet (such as a soundproofed recording room or an anechoic chamber), the brain tends to "fill in" by supplying sounds that aren't actually there.

 URINE DOESN'T HELP WITH JELLYFISH STINGS

In fact, peeing on a jellyfish sting may provoke the nematocysts—the toxin-containing cells that are left on your skin—to release additional poison. Rinsing the skin affected with seawater is more effective for pain relief.

 THE FIRST EFFECTIVE CURE FOR SYPHILIS WAS FULL OF ARSENIC

Arsenaphenamine was the active ingredient in the successful product Salvarsan, which went on sale in 1910. Its developer, the German biochemist Paul Ehrlich, also coined the term "magic bullet" to describe it.

 RAMSES II HAD A VALID PASSPORT 3,000 YEARS AFTER HE DIED

In 1974, the mummified remains of pharaoh Rameses II traveled from Egypt to France to undergo special preservation treatment; the passport issued for him had an expiration date of 1981, giving him six years to return home.

 ACTINIUM-225 MAY BE THE RAREST DRUG IN THE WORLD

It's a radioisotope that occurs naturally, but only in tiny amounts (only enough to treat around 1,700 patients in total). It's unrivaled, though, at treating late-stage prostate cancer. Drug companies are now exploring ways in which it can be produced artificially.

87 WHEN WAS PLASTIC SURGERY INVENTED?

Today we tend to think of plastic surgery as a modern art; the immediate connection you might make with the term would be "improvement" surgery, such as breast or buttock enhancement, or facelifts. However, it was first developed in ancient times to repair wound damage or the ravages of disease.

FIRST DAYS

The first mention of plastic surgery comes in the *Sushruta Samhita*, by the Indian healer Sushruta, written in the sixth century BCE. Among a good deal of information about healing plants and the treatment of diseases, he described both skin grafting (the technique of "patching" wounds with skin taken from other parts of the body) and surgery to reconstruct the nose—the first rhinoplasty, in which a flap of skin from the forehead was used to graft onto a damaged nose.

Seven hundred years later, the Roman physician Aulus Cornelius Celsus covered grafting and reconstruction in *De Medicina,* the surviving fragment of a much larger work.

STEP-BY-STEP SURGERY

Perhaps the first pioneering European surgeon was Antonio Branca, himself the son of a Sicilian surgeon, who is credited with inventing the "Italian method" of skin grafting and rhinoplasty. A German, Heinrich von Pfolspeundt, recorded Branca's method in fascinating detail in his *Buch der Bündth-Ertznei*, published in 1460. First the surgeon made a kind of template for the nose in either parchment or leather; this was laid down on the patient's forearm and traced around. The shape was cut around to create a flap of skin, but left attached to the forearm at the lower edge, the forearm raised to the face, and the free portion of the flap stitched in place over the nose. The graft was then left for ten days to "take," the patient spending the time with their arm strapped to their face so the graft could not be pulled away. After this time, the lower flap was cut through, so that the arm could be freed and the final, lower part of the graft stitched in place around the nostrils. The huge advance in Branca's method was that it ensured the skin was "alive" by leaving it attached to its original site while the graft took place—although it must have been very stressful for the patient.

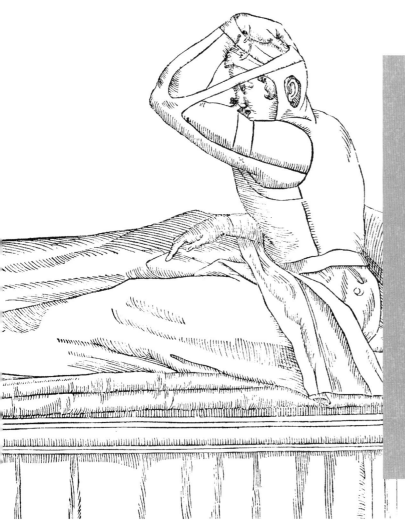

WHY "PLASTIC" SURGERY?

The term was invented by Pierre-Joseph Desault, a surgeon and teaching professor at La Charité and Hôtel-Dieu hospitals in Paris in the second half of the eighteenth century. He was the first to use the term *plastikos*, from the Greek, meaning "able to be molded," in connection with the operations, usually skin grafts, that aimed to correct or improve scars and facial deformities. In allowing large numbers of students to attend operations, he created a new generation of surgeons who followed his innovative methods and techniques.

88 WHAT'S THE BEST WAY TO PRESERVE A HUMAN BODY?

The best method of preservation depends on what use your body is to be put to when you're dead. Are you aiming to turn up to your religion's specific afterlife in good shape? Or do you want to explore the possibility of being brought back to life when humans have discovered how to defeat death?

WELL-PRESERVED

Mummification has been practiced for millennia, most famously by the ancient Egyptians. If you want to go to the afterlife with a well-preserved exterior it's a good choice, but for top levels of preservation your interior organs will need to be removed. After you've been eviscerated, things improve: Your body will be rinsed with wine, adorned with spices, and left in a thick layer of salts to dry out. For organ retention, a natural mummification in a peat bog may be the best option. The remains of Tollund Man, found in Denmark in 1950, were so well preserved that the peat cutters who found him thought at first that his body—which had been interred for around 2,000 years—was a recent burial. Unlike the desiccated mummies of ancient Egypt, peat bodies look more recognizable, their soft tissues as well as their skin and bones preserved in an extremely acid environment that contains hardly any oxygen.

EMBALMING

Vladimir Lenin is probably the poster boy for this—well over ninety years after his death, the Russian leader is looking good (in fact, his remains have been constantly improved with an annual fresh embalming and cosmetic overhaul). Preparations for the public display of his body involve lengthy baths of, among other ingredients, formaldehyde, glycerol, and hydrogen peroxide. Some pieces have quietly been replaced: Artificial hair and eyelashes have been added, and when the body is displayed in public, his clothes go over a thin rubber "coat" to keep the chemicals close to his skin.

FROZEN

Not the Disney hit, but the science of cryogenics: freezing dead bodies in the hope that they can be brought back to life in the future. As soon as you die, you'll be immersed in an ice bath to get your temperature down as rapidly as possible. Next, your blood will be drained and replaced with antifreeze before you're packed in ice and cooled further over the next fortnight until you reach the magic temperature of −320°F, after which you'll be stored in liquid nitrogen until scientific advances allow you to be either brought back to life or cloned.

SWEET RELEASE

No preservation technique is odder than mellification—the practice of preserving bodies in honey. The technique was recorded in the *Bencao Gangmu* (*Compendium of Materia Medica*), an extensive Chinese collection of medicines authored by the sixteenth-century apothecary Li Shizhen. Among the cures listed is one for broken bones: The sufferer should eat small pieces of mellified bodies—a sort of cannibalistic *marrons glacés*. As recorded, mellification was a spectacularly lengthy process that began while the subjects, usually elderly holy men, were still alive, by feeding them exclusively with honey. When they died (and diabetes must have been on the cards, given the sugar content of their diet), their bodies were immersed in huge jars full of even more honey and left there for a century. Only then were they removed and broken into small—and very pricey—pieces.

89 WHICH IS THE RAREST BLOOD TYPE?

Usually the question "Which blood type are you?" is only asked in the context of someone needing a blood transfusion—but until comparatively recently, all blood was believed to have more or less the same qualities. It was only when blood was classified into different types that transfusion ceased to be an often-deadly lottery.

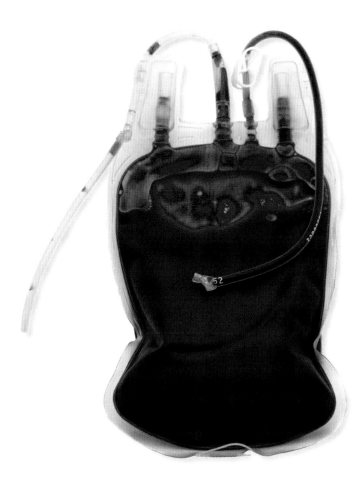

PUTTING BLOOD BACK

The idea of "putting blood back" into a body that had bled heavily through injury or accident is an old one; it was first tried in the seventeenth century, shortly after the English physician William Harvey had discovered circulation. By the beginning of the nineteenth century, transfusions were regularly attempted, and sometimes they worked; at other times, though, patients had catastrophic reactions and the failure rate was still high.

THE DISCOVERY OF BLOOD TYPES?

At the beginning of the twentieth century, Karl Landsteiner, a pathologist working at the University of Vienna, Austria, discovered the existence of blood types (he would later be awarded a Nobel Prize for his work). First he identified the A, B, and O groups, and a few years later his colleagues found the AB group. Blood cells can feature many different antigens, that is, substances that prompt the production of antibodies in the immune system. And at the end of the 1930s it was discovered that the presence or absence of a particularly powerful antigen, called the rH factor, could also affect your blood type—if you had the factor, you were positive, if not, you were negative. It's at this point that most people's knowledge of blood types ends: They know that O is the "universal" blood type, and that within each category there are "plus" or "minus" groups, but not really what that means.

IT TAKES ALL TYPES

Which group you belong to is crucial when it comes to either giving blood or having a transfusion. Someone given blood from the wrong group can have bad, sometimes very bad, problems. The blood may clot or coagulate, or the body may reject it altogether, leading to a catastrophic reaction. And even today, small, even tiny blood subgroups in which particular antigens are either present or absent are still being discovered. These are sometimes found in people who have a smaller gene pool as a result of being members of small societal groups.

These small blood groups can create problems when it comes to a transfusion. "Bombay blood," for example, also known as the HH group, is so called because it was first written about in Mumbai in 1952 when doctors found two patients for whom no available blood in transfusion "worked." Eventually they were found to be unable to express the H antigen, which is present in all common blood types. This meant that although they could give blood to any of the other groups, they could only receive blood from a fellow member of the HH group.

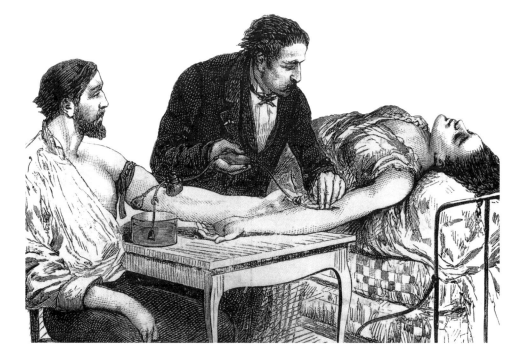

90 DO PIERCINGS HAVE HEALTH BENEFITS?

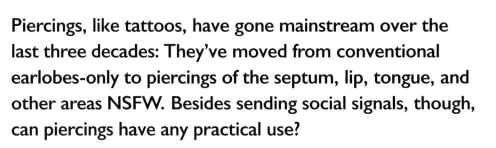

Piercings, like tattoos, have gone mainstream over the last three decades: They've moved from conventional earlobes-only to piercings of the septum, lip, tongue, and other areas NSFW. Besides sending social signals, though, can piercings have any practical use?

PRESSURE POINTS

In Ayurvedic medicine, specific piercings are believed to help with various health problems. Many correspond with the pressure points of Chinese acupressure and acupuncture, tapping into the energy pathways in the body to maximize their efficiency. The traditional earlobe piercing, though, carried out on small babies, sometimes at just a few days old, is probably judged the most important. The *Sushruta Samhita*, an encyclopedia of health written in the sixth century BCE, sets out the rituals for the Karnavedha, the double ear-piercing ceremony, which is one of the sixteen samskaras—a word that means something like "rites of passage." It also lists all the good things it will do for a child's health. Read through them, and you'll see why, as an all-around protective health measure, the Karnavedha is key. Boys have their right ears pierced first, then the left; for girls, it's the other way around.

KNOW YOUR MERIDIANS

Earlobes are thought to hold important meridian points that link with both sides of the brain. It's held that piercing them will help a baby's brain to develop properly and aid with memory function all through life. There is also a point in the middle of the lobe that links with the reproductive system and, for girls, an accurate piercing will ensure that their menstrual cycles stay healthy. Yet another point aids digestive function.

QUIZ

MEDICINE

Is there a hole in your head where the facts from this chapter fell out, or has everything stayed inside your brain? Take this quiz to find out.

QUESTIONS:

1. Who discovered blood types?

2. Where is the first mention of plastic surgery found?

3. When was the last death from smallpox?

4. What makes superbugs difficult to treat?

5. What's the difference between a trepanation and a craniotomy?

6. If you wanted your body to be cryogenically preserved, what temperature would you be kept at?

7. What is mollification?

8. What can you do if you have rare, HH group blood?

9. What percentage of Europe's population did the Black Death kill in the fourteenth century?

10. In Ayurvedic medicine, what is the name of the ear-piercing ceremony?

Turn to page 248 for the answers.

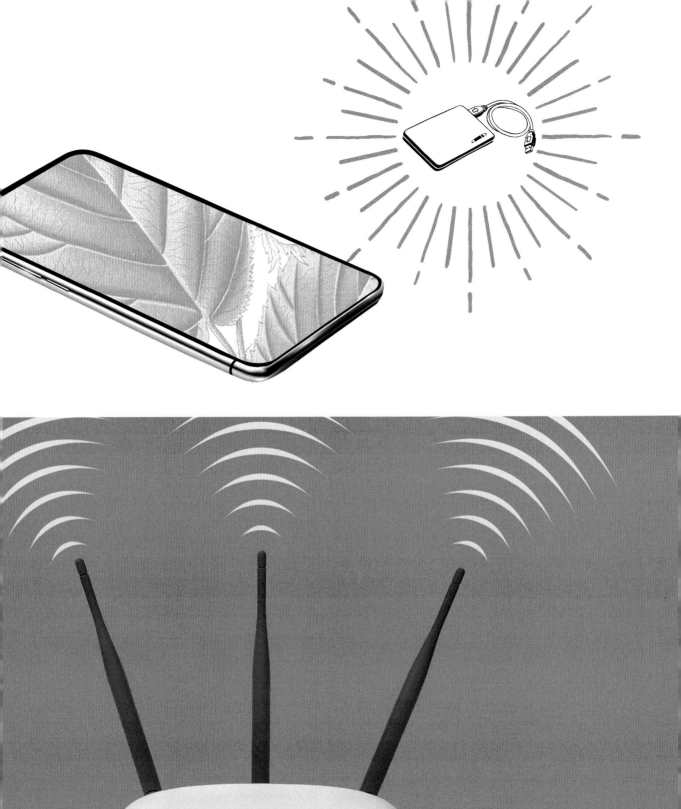

COMPUTING AND TECHNOLOGY

91 HOW DOES WI-FI WORK?

Wi-Fi (which is a marketing term that doesn't stand for anything) is absolutely everywhere these days; homes, offices, cafés, and even some entire cities have Wi-Fi. All Wi-Fi works by transmitting information-carrying radio signals between an internet-connected hub and the Wi-Fi–enabled device.

CREATING RADIO WAVES

Before we look at what makes Wi-Fi waves special, we need to consider how radio waves work normally. Radio waves are very easy to make; you just need a simple antenna. Radio waves are a type of electromagnetic radiation (just like light) and they can be created by simply moving some kind of electrically charged particle, like an electron. In a radio antenna, electrons move up and down a metal pole to create radio waves. By moving them up and down at different speeds, radio waves of different frequencies can be made. Anything that is powered by electricity, which involves the movement of electrons, also gives off radio waves; since most modern technology is powered by electricity, there are lots of radio waves around you all the time. Radio antennae, however, are specially designed to generate a single, large, and clear radio wave that will be able to travel long distances.

CARRYING INFORMATION

We often picture radio waves as simple regular patterns, but in order to be useful they must be able to carry information somehow. But how? To transmit information, the normal radio wave frequency (for Wi-Fi this is usually 2.4 GHz) is "enveloped" by an information signal. This means that the size of the wave is squished into the shape of the information signal before being broadcast out. This is like taking a stiff wire with just a little bit of give in it and bending it into a shape; the main curve is still the same but the finer details contain the information. When the signal then reaches a device such as a phone or a laptop, the information can be extracted from the shape of the radio signal. This allows for rapid data transfer, which makes wireless internet usage possible.

IS WI-FI DANGEROUS?

Some people are concerned about the possible health risks associated with living in an environment that is constantly full of radio waves. However, on this topic the science is conclusive. There is no danger from the radio signals put out by Wi-Fi devices, mobile phones, telephone poles, or any other modern electronic equipment. This is not only because the radiation from them is non-ionizing and thus safe but also because any amount we produce through our technology is massively dwarfed by the colossal amount of radiation coming from space that passes through us without harm every day.

92 WHY ARE FIBER OPTIC CABLES THE FUTURE?

Since the days of the telegram, information has been transmitted through copper cabling. From the transatlantic cable to phone lines in homes, copper has been the material of choice. More recently, however, fiber optic cables are becoming more common, as they are able to carry much more information more reliably.

UNDERGROUND COPPER CABLES

Information signals can be sent along copper cables in the form of pulses of electricity. In the early 1800s, this was originally just simple on or off signals used for Morse code in telegrams, where a combination of long or short pulses can be converted into letters. As things became more advanced, it was possible to send more than a single signal down the same wire by varying the intensity and phase of the current. These copper cables were often bundled together in some protective shielding and would be laid down underground, where they would be kept safe from accidental damage and external forms of interference.

WHAT ARE FIBER OPTIC CABLES?

You may well have come across fiber optic cables before: They are a staple of stadium flashing wands and ultra-modern lighting. They consist of a very simple glass or transparent plastic tube through which light can be transmitted from one end to the other, even as they move and bend. A strand of optical fibers allows light to move down the tube without leaking out. This is due to total internal reflection—the nature of the material used for the optical fibers means that any light that hits the side of the tube bounces off and back into the fiber. A dark cladding is put around the glass to help with this process.

DOING A BETTER JOB

Fiber optic cables work using pulses of light. Like the original copper cables, they send 1s and 0s as on or off signals from one end to the other. The difference, however, is that it is possible to vary the frequency of light being sent to allow for at least ten times the number of signals as even the most advanced copper cabling. It's not just information carrying where fiber optics win out. They have better transmission length, meaning that they are able to carry information over a much longer distance without the data starting to degrade.

IS THE FUTURE FIBER?

They are much more durable than their copper counterparts, as well as longer lasting. Even protected copper cabling can oxidize (meaning that exposure to air changes the material) over time, which causes it to be less efficient. Furthermore, as copper cables use electricity, it is possible for them to be affected by external fields that can damage their signal, whereas fiber optic cables avoid this problem by using light. Fiber optic cables are also more secure: It is impossible to tap fiber optic cables and they don't leak out signals like copper does. The only reason fiber optics aren't everywhere at the moment is that they're more expensive and existing networks of copper cables are already in the ground. However, as copper gets more pricey and fiber optics become cheaper, as the old cables need to be replaced it seems the future is in fiber.

93 WHAT DO TRANSFORMERS DO?

Transformers are everywhere, from the roadside building plastered in warning signs to the strange box on the chargers for all your gadgets. Despite their ubiquity, however, you may not even know what they're for. Transformers increase or reduce the voltage of electricity ready for use or transmission.

WHAT GOES ON INSIDE THE BOX

A transformer consists of three main elements: the incoming primary wire, the outgoing secondary wire, and a magnetic core. When the primary wire has a current applied to it, this induces a magnetic field in the core which then, in turn, induces an electrical current in the secondary wire. This process allows for power to be transferred from one side of the transformer to the other. What changes across the transformer is the electricity's voltage; how it changes is dependent on the number of coils of wire on either side. If the secondary wire has more coils than the primary wire, then the voltage is increased; if the reverse is true, then the voltage is decreased.

THE NATIONAL GRID

In some countries, power plants produce electricity at 25,000 volts, but this needs to be changed before transmission. As electricity travels along overhead lines, the lower the current is the less power is lost, and so the voltage is increased to 400,000 volts. This voltage of electricity would blow up all the electronic devices in homes, so before it reaches them it goes to small local substations where it is reduced to 120 volts (known as household power). This power is then sent to houses and businesses where consumers can plug in devices that are designed to work at this voltage or that have their own small transformers (usually as part of the power cable) to reduce the voltage down to the correct level.

FIVE QUICK FACTS

THE PROTOTYPE FOR THE COMPUTER MOUSE WAS MADE OF WOOD

Developed by Doug Engelbart at the Stanford Research Institute, it was named for the "tail" (the flex) that emerged from the top back. Patented on November 17, 1970, it has been in use ever since.

IN THE FUTURE, CLEANING YOUR TEETH MAY TAKE JUST TEN SECONDS

We've long been told that a proper clean should take a hardworking two minutes, but new toothbrush designs aim to "pulse" your teeth clean in just a twelfth of that time.

THE WORLD'S FIRST COMPUTER WEIGHED 30 TONS AND FILLED A LARGE ROOM

Computers have become progressively smaller since ENIAC, the world's first general-purpose digital computer was made in 1945. Today, the world's tiniest computer is dwarfed by a grain of rice, and they're still shrinking.

SPIDER SILK IS ONE OF THE HARDEST MATERIALS TO IMITATE

Scientists have been trying to replicate the strength and flexibility of spider silk for decades. In 2021, the University of Washington announced that it had made an artificial material that matched and outperformed the natural version.

IF YOU'VE EVER WONDERED WHAT IT'S LIKE TO HAVE THE EARS OF A CAT, YOU CAN BUY THEM

In 2012, a Japanese company manufactured sensitive cat ears. They look like a cute kawaii headband, but they're sensitive to brain waves, programmed to respond to the moods of the wearer by changing position.

94 HOW DO COMPUTERS STORE 1S AND 0S?

We use computers for nearly everything without ever really knowing how they work, or what the physical components look like. While we generally understand that computers run off 1s and 0s, how this actually happens is a complete mystery to many. Computers physically store 1s and 0s inside magnetic domains.

MAGNETIC BITS

Computer hard drives are made out of disks that are able to rotate, and the information stored on them can be read or written using a head in a process that works like an old record player. In record players, the information is stored in the grooves in the record that the needle "reads" as it passes over, but in hard drives it's encoded in tiny sections of the disk called magnetic bits. (A byte is made up of eight bits.) These magnetic bits are made out of several even smaller "grains," which are naturally occurring structures in metals. These grains act like tiny magnets, and a magnetic bit can be made by linking some of these grains together to act as one bigger magnet.

It is possible to store data using these magnetic bits. An individual magnetic bit can be pointing left or right (or up and down, depending on how it's built). When you want to access the information, a read head is passed over the top of the bits. When it detects that there has been a change of direction from one domain to the next, it reads that as a 1. If there has been no change, it instead reads a 0. A hard drive can either read or write (by flipping the magnets) millions of bits every second.

THE FUTURE OF STORAGE

There is a constant drive forward for more storage on hard drives and for them to become even smaller as technology changes. One of the types of hard drive just on the horizon is heat-assisted magnetic recording (HAMR). An obvious way to make a hard drive fit more data onto it is to make the bits and grains smaller. However, when you make them smaller they become more susceptible to environmental influences that can cause the magnetic direction of the bit to change randomly, which would corrupt the data. To combat this, a magnetic layer with a higher coercivity can be used; this means that the magnetic directions are more strongly held in place. However, when the higher-coercivity material is used, it also

SOLID STATE DRIVES

Instead of magnetic materials, solid state drives (SSDs) use electrical transistors that are either open, allowing electrons to flow through (a 1), or off, stopping them (a 0). This allows for the information to be recorded and written faster and makes the device more durable, as there are no mechanical parts to break down. However, SSDs are more expensive and store less data than traditional hard drives.

makes it harder to write new data onto the hard drive, which slows down the computer. In HAMR, before trying to write new data, a laser is used to heat up the material, which lowers its coercivity. The information is then written onto the hard drive, which then cools, locking it into place. This means that more data can be stored in the same space.

95 HOW IS A POWER PLANT LIKE A KETTLE?

Keeping the lights on is no easy feat. Enormous buildings are dedicated to producing enough energy for all our daily needs. They may seem incredibly advanced, but power plants have something in common with a very ordinary household object. While it's for very different reasons, power plants and kettles both produce a lot of steam.

GENERATORS

It is possible to induce an electrical current in a wire by moving it in a magnetic field. The most efficient way to generate electricity by this method is to rotate a coil of wire within a magnetic field. Doing so at a constant rate provides a stable current of electricity passing out of the ends of the wire. As the wire will be traveling through the field one way for half a turn (up) and then the other for the second half of the turn (down), this causes the electrons in the wire to flow one way and then the other. This is known as alternating current (AC) and is the normal type of electricity you find in your home.

GIANT KETTLES

While it is possible to generate alternating current using a simple hand crank or wheel to turn a coil of wire, at a power plant huge amounts of electricity must be generated, so a much more efficient way of turning a coil of wire is needed. Almost all power plants do this in the same way: Some fuel is burned and the resulting heat is used to boil a vast vat of water like a kettle. Oil, gas, coal, and even nuclear power plants all use their respective fuels to heat up the water. Once the water reaches boiling point, a superhot steam (up to 1,100°F whereas a standard kettle only produces steam at 212°F) rises rapidly and passes through a series of turbines, causing them to spin. These spinning turbines are hooked up to generators that are used to produce the electricity that is sent across the country.

RENEWABLE ENERGY

Most renewable forms of energy also rely on the rotation of a turbine to generate electricity. In a wind farm, it is obviously the wind driving the turbine, while in hydroelectric power plants the turbines are turned by large volumes of water falling from a dam. Even in plants that use biomass and geothermal energy, that energy is used to boil water for the steam that can then turn the turbine. The exception is solar power. There are some solar power plants that use mirrors to reflect the Sun's rays into the plant to heat water and create steam. However, it is more common that the photovoltaic cells in solar panels absorb sunlight and are then able to release electrons, which creates the desired flow of electricity.

FUSION: FUEL OF THE FUTURE!

Scientists have billed fusion energy as a promising and bountiful energy source in the next few decades. Fusion replicates the conditions of the Sun's core and is able to force two hydrogen atoms to fuse together into a single helium atom, a process that releases 10,000,000 times more energy per pound than any fossil fuel. But even when fueled by this advanced form of energy, power plants will still use the energy to heat up water, to create steam to turn a turbine.

96 WHAT MAKES A QUANTUM COMPUTER SPECIAL?

Computers are always getting smaller and faster, but we're starting to reach the limit. Transistors (the smallest part in a computer) are already about 500 times smaller than a red blood cell; at this level, quantum effects mean that traditional components just stop working. What makes quantum computers special is that they take advantage of quantum effects to store and process huge amounts of information.

QUBITS

Information in normal computers is stored as a bit, as part of a magnetic medium (see page 212). A bit can have two values, either 1 or 0. A qubit, on the other hand, is any property of a material that can have only two possible states, such as the spin of an electron (up and down) or phase of a light photon (horizontal and vertical). The qubit can use the quantum effect of superposition to be a bit of both at the same time, and it only becomes one or the other when you use it. This means that you can store multiple pieces of information in the same space. Old 8-bit video games used 8 bits to store a single piece of information, but if they were to use 8 qubits they could store 256 instead because the qubits can be all of the possibilities of 1s and 0s at the same time. This number grows exponentially, too: Many modern computers use 64 bits to store information, meaning that an equivalent qubit computer could store 18,446,744,073,709,551,616 times the information!

WHY SHOULD I CARE?

While you shouldn't expect to see your home desktop or mobile phone become quantum-based anytime soon, just because quantum computers are at the early developmental stage doesn't mean they won't soon be affecting your life. The increased power and storage of quantum computers will be used by scientists to create even more accurate models and work on more complex problems, vastly improving research of all kinds. More immediately concerning, however, is the implication for data security. Modern computers use complicated mathematical codes to keep us safe. A modern supercomputer would take about a billion times longer than the existence of the universe to crack a standard 128-bit advanced encryption standard (AES) key, but a quantum computer may be able to do it in just a few moments. This is leading security experts to already reconsider how to keep us secure in the future.

SPEED READING

It's not just in storage where qubits shine. It may also be possible to use quantum entanglement to link two qubits together so that when one is a 0 the other is always a 1, or so that when the first is a 1 the other is also a 1, or really whatever combination you choose. This would mean that in reading a line of qubits you'd know what two lines say, cutting the time taken to read them in half and thereby making processing incredibly fast!

97 WHAT IS A PIXEL?

We look at pixels all the time on our computers, phones, and other devices, but do we really know what they are or how they work? A pixel (short for picture element) is the smallest element of a digital screen; these combine to create images.

PAINTING A PICTURE

Pixels are simple squares that can be made to be any color—but only one at a time. Much like in a mosaic, it is possible to then put many of these pixels together to begin to form an image. With just a few pixels it is only possible to make simple shapes, but as you increase the number of pixels the image begins to look more and more realistic. With a high enough number of pixels, it becomes impossible to distinguish them at all, only the completed image.

THE LIGHT FANTASTIC

The human brain is very easy to fool, especially when it comes to our vision. Pixels don't actually need to be able to produce the hundreds of thousands of colors we see, just three: red, green, and blue. Much in the same way that printers only need three inks and are able to mix them together into all the different colors you could want, pixels are able to use varying amounts of the three base colors to produce a dazzling array. Different types

of screens work in slightly different ways, but one of the most common types is the liquid crystal display (LCD). A pixel on an LCD screen is made up of three subpixels: a red, green, and blue one. These are then covered by three colored filters, one for each color. Initially, the crystals in each of the filters are closely knitted together and block out most of the light from all of the subpixels. However, when an electrical charge is applied to a filter the crystals move apart, allowing more light to pass through. By carefully controlling the three different filters it is possible to generate literally millions of different colors, which can be used to create realistic images.

RESOLUTION

Lots of people, especially advertisers, will tell you that when it comes to pixels, more is better. You may know some values for resolutions even if you don't know what they really mean. A common resolution for computer screens is 1,920 × 1,080; what this means is that from left to right there are 1,920 pixels and 1,080 from top to

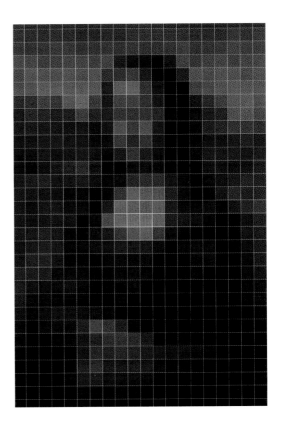

bottom, for a total of 2,073,600 pixels. The largest phone screens can be as pixel-heavy as 1,440 × 2,560, giving a whopping 3,686,400 pixels in all. It is also worth noting that a resolution only tells you how many pixels there are in a given space, not how big that space is. So, for example, a 1,080 × 1,920 phone screen will have a lot more pixels in the same space than a 1,920 × 1,080 computer screen, leading to a higher-quality image.

98 HOW GREEN ARE ELECTRIC CARS?

There were more than 10 million of them on the world's highways at the end of 2020, and many countries have a tight time commitment to phase out gasoline-powered vehicles. So how green are electric cars, and will they get greener?

ROOM FOR IMPROVEMENT

Taking all factors into account, it's estimated that today's electric car is around 50 percent better in environmental terms than its gas-powered equivalent—and that's because it doesn't emit CO_2 when driven. The reason it's not 100 percent is due to negative environmental impact at three other points: the manufacture—specifically of the battery; the "greenness" or otherwise of the power that charges it; and the ease of recycling the battery when the car's life is over.

Electric car batteries rely on a number of earth metals—nickel, manganese, cobalt, and lithium—that are mined at notoriously high human and environmental cost. There's also the question of the electricity used to charge the car when it's in use. Unless this is clean-sourced, it adds to the overall environmental impact. Finally, the batteries in electric cars are harder to recycle than those in gas-powered cars; the lithium content in particular is hard to extract and recycle. This will improve as electric cars become the norm, and alternative afterlives for batteries may also be found. A 2020 study by the Massachusetts Institute of Technology suggested that after exhausting their auto use the batteries could enjoy a further decade of useful life as backup storage for solar power systems.

The verdict? Electric cars are greener than gas-powered cars, but perhaps not as much greener as you might have assumed. Some aspects of their use and manufacture will become more environmentally friendly as the market grows, but trains or bicycles are still greener ways of getting around.

QUIZ
COMPUTING AND TECHNOLOGY

The information superhighway of your brain should be able to remember everything in this chapter, so power through and see what you know.

QUESTIONS:

1. What does HAMR stand for?

2. In a computer context, what are "grains"?

3. How many colors do pixels need to produce an image?

4. How many elements does a transformer have?

5. How much better is today's electric car than its gas-powered equivalent?

6. What is a qubit?

7. How are fiber-optic cables superior to the old copper variety?

8. Why isn't the radiation from Wi-Fi bad for you?

9. How many electric cars are there globally?

10. What does fusion energy do?

Turn to page 249 for the answers.

THE FUTURE

99 COULD A CLOAK OF INVISIBILITY BECOME A REALITY?

Being able to disappear at will has been close to the top of everyone's list of desirable superpowers for decades. But making things invisible to the naked eye has proved much harder to do in real life.

BENDING LIGHT FREQUENCIES

Everyday light, the means by which we see things, is made up of a full range of waves of different frequencies, with each color having its own range. The way we see something specific depends on which of the waves it absorbs. A red ball, for example, reflects the red waves back to us while absorbing all the other colors, so that we see it as red. Most experiments in invisibility until recently have relied on meta-materials, man-made fabrics that split light waves, causing the light to go around objects—meaning that the light couldn't "see" them. The results did achieve a degree of "invisibility" but left the viewer with an effect that was clearly visually odd.

SPECTRAL CLOAKING

In theory, if an object absorbed all the light waves of every color, you wouldn't be able to see it. To see right through the object, though—that is, for the object to be invisible, and for the viewer to see what's behind it—you'd have to manipulate the light waves so that they reverted to their normal behavior after they had passed through it. In 2018, a team from the National Institute for Scientific Research in Montreal developed an experiment that moved things closer to "real" invisibility by moving colors onto the frequency of different light waves, and thus colors, as they approached the object.

To disguise the red ball, for example, this would mean that the red waves were shifted over into the frequency of other colors, so that the red wasn't seen. Once the waves had passed through the object, they would be returned to their original frequencies, so that the space behind the object was visually "filled in" and the object, in visual terms at least, ceased to exist.

IT WORKS IN PRINCIPLE

The technique has been christened "spectral cloaking," and in trials it worked, although so far only in a limited way—it disguised in one dimension only, meaning that an observer would have to be looking in a single, immovable direction for the object to disappear. And it worked on a small optical filter, not (so far) on any kind of larger everyday object. Such constraints mean that the current spectral cloak wouldn't be useful in fantasy wizarding applications, but the experiment did prove the principle is workable. It's a promising start.

100 HOW CLOSE ARE WE TO EATING LAB-GROWN MEAT?

There's already a proven market for vegetarian or vegan products that mimic meat—in some cases almost eerily successfully. But the idea of "real" meat that can be grown, cruelty-free, in a lab—and eaten without guilt—has attracted a lot of attention (and investment) over the last decade.

GROW YOUR OWN

Although so-called lab meat hardly exists in any commercially consumable form yet—just a small quantity of chicken nuggets, bioengineered by a San Francisco–based company, were cleared for sale in Singapore in 2021—the speculative market is already estimated at a couple of hundred million dollars annual turnover and has been projected to nearly triple by 2025.

There's a strong counterargument to lab meat optimism: it may prove to be too difficult and too expensive to produce in any quantity. Lab-grown meat has featured, in minute amounts, at press events attended by chefs and influencers, but the focus has tended to be on what it tastes like, and how similar it is to meat from farmed animals, rather than on the practical aspects of exactly how it could be produced at scale. Doubters point out that culturing meat cells en masse calls for facilities that are both huge and unfeasibly clean—more like pharmaceutical factories than any food-production model. Cultured cells are "fastidious" and tricky to grow (although avian cells have proved more amenable than bovine ones, which is why it's proved easier to "make" turkey and chicken than beef). They also don't have any built-in immunity, so the slightest impurity in conditions could wipe out whole colonies easily. You may not be picking up a manufactured steak at your supermarket anytime soon.

YOU ARE WHAT YOU EAT

Once the idea of cruelty-free meat took hold, it was inevitable that someone would look at the possibility of meat made from human cells. The neatly titled *Ouroboros Steak Kit* (named for the mythical serpent that consumes its own tail) featured as part of an exhibition, "Designs for Different Futures," in 2019–20, although it was developed as an art project—offering people food for thought rather than immediate consumption. It posed the idea that human cells, harvested from a cheek swab, could be applied to a "scaffold"—a supplied framework made from fungal mycelium cells—and fed with a serum derived from donated blood. Given favorable conditions, the cells would, in theory, take around three months to grow into a juicy steak, which could be consumed by its maker without guilt: "you are what you eat" taken to literal extremes. The *Ouroboros Steak* was created by a team—an artist, a biophysicist, and an industrial designer—and was intended to be thought-provoking. Although it did attract a number of queries from would-be buyers keen to grow their own meat, it also resulted in a negative social media storm when it was exhibited. Grace Knight, the industrial designer on the project, responding to comments, described the project as "technically not" cannibalism.

101 CAN PLANTS THINK?

In one of his last works, *The Power of Movement in Plants*, written with his son Francis and published in 1880, Charles Darwin suggested the root-brain hypothesis: the idea that plants have a brainlike structure in their roots. Over ninety years later, in 1973, another book, *The Secret Life of Plants*, developed the basic idea more, making extraordinarily strong claims for the consciousness of plants.

GREEN THINKING

Darwin had begun a debate about plant intelligence that is still ongoing. Plants can do many remarkable things; the question he had raised was whether or not they do them with any kind of consciousness. Those who believe they do frame plant behavior as plant "neurobiology." Those who don't (including most of the conventional scientific establishment) point out that plants lack a nervous system, so studying them as though they have one doesn't make sense. This isn't to downplay the extraordinary qualities of plants but to give them their own space in studies rather than needing to anthropocentrically compare them with our own abilities. "Animals learn, plants adapt" goes the thinking.

ADAPTATION AND REACTION

Unlike animals, plants can't move, so in order to survive they have adapted to their immediate environments in many ways. They can lose a very high proportion of their body and still regenerate (for example, if something starts to eat them, they can't move away), and they have sophisticated biochemical abilities, both reacting to chemical molecules around them and producing them for themselves. In one experiment, an *Arabidopsis* mustard plant was played the vibrations of a cabbage white caterpillar—one of its predators—eating; in response, it flooded its leaves with glucosinolates—chemicals that made them too bitter to be palatable.

Does this sort of reaction indicate that plants can think? Probably not in any way we would recognize as conscious thought. But that may not really be the relevant question: given all the ways in which plants are remarkable enough in their own right, perhaps it's best to explore their many different adaptations and abilities without crediting them with full animal-type thought processes too.

PSYCHIC PLANTS

Perhaps one of the strangest stories in the study of plant thought is that of Cleve Backster. Backster was a CIA polygraph expert who, in the mid-1960s, began to experiment on plant cognition, and was soon crediting plants with extraordinary powers. He hooked up plants to polygraphs, claimed to be thinking of harming them or burning them, then recorded them reacting with acute alarm in the test results. The idea that plants were literally mind readers became popular with a whole range of alternative thinkers, from hippies to Scientologists, and Backster published a number of articles in the *Journal of Parapsychology*, and some successful books. However, his findings could never be duplicated in experiments carried out by others, and his beliefs were ultimately rejected by serious scientists. He died in 2013, loyal to his ideas to the end.

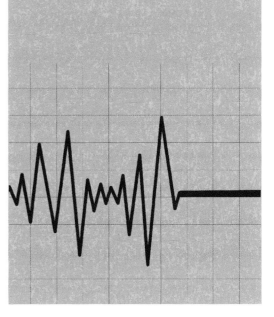

102 WHAT DO THE WORLD'S SMALLEST COMPUTERS DO?

Computers have gone from small to tiny to absolutely minute over the last decade, and they still seem to be shrinking. But how are such miniature machines deployed?

CHIPS OFF THE BLOCK

The University of Michigan has form when it comes to small computers. Teams there were responsible for the creation of the Michigan Micro Mote, which measured 4 by 2 millimeters. But in 2018, they announced a new model, less than one-third of a millimeter square, making the Micro Mote look clumsy! Many of the likely applications for this new computer are medical. It was designed to act as a temperature sensor, so small that it can trace raised temperatures in small cell clusters that may prove cancerous, but such a tiny device could also monitor pressure if implanted in the eye of someone suffering from glaucoma, for example.

OUT IN THE FIELD

The scale has already proved useful for naturalists. An adapted version of Michigan's Mote had recent success as part of a field study of some very small snails. Naturalists wanted to know why one species of tree snail (*Partula hyalina*)—native to the Society Islands in the South Pacific—had survived while other tree snails had been predated by species introduced to their habitat, and ultimately gone extinct. As a protected species, the snails couldn't carry sensors directly, but the adapted computers were glued to the leaves in which they rested instead, collecting valuable information while the snails slept.

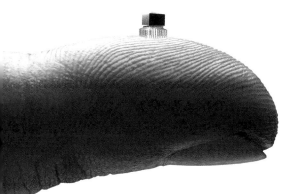

FIVE QUICK FACTS

 AROUND 6,000 COMPUTER VIRUSES ARE RELEASED EVERY MONTH

Originally developed as piracy guards for new software, computer viruses are now one of the more irritating elements of digital life.

 IN THE FUTURE, AMPUTEES MAY BE FITTED WITH EXOSKELETONS

Already in development, the devices wrap around the wearers, then use microprocessors and electric motors to "power" movement.

 ANCIENT VIRUSES MAY BE RELEASED AS THE GLACIERS MELT

Scientists in western China have identified previously unknown viruses, believed to be more than 15,000 years old, in ice cores of ancient glaciers.

 LONG PERIODS OF VIRTUAL REALITY GAMING CAN ALTER YOUR PERCEPTION OF TIME

If, when gaming, you've had the sensation of hours passing in a flash, you weren't wrong. Too much time in virtual reality warps your perception of time—and the phenomenon now has an official name: "time compression."

 SQUID MEMBRANE MAY BE USED TO MAKE CAMOUFLAGE IN THE FUTURE

"Smart skin" camouflage may learn from the squid's extraordinary ability not only to change color but also to vary the amount of light it reflects.

103 CAN YOU BUILD WITH MUSHROOMS?

Today, mushrooms are being pressed into all kinds of roles, way beyond the culinary. They've been repurposed as a packaging material and insulation, "grown" into mushroom leather, and now they're being looked at for a role in the construction industry. The proposal is that they can be grown into bricks.

BIOLOGICAL BRICKS

Construction has a huge carbon footprint: it's responsible for 10 percent of global CO_2 emissions annually. So any material that could help to cut that total without having other major disadvantages is bound to raise interest. But are mushrooms tough enough to build with?

Enthusiasts believe they are. Mushrooms have a lot of pluses. They're carbon-neutral and fire-resistant, and when their life as bricks is over, they don't pose an expensive disposal problem—they can simply be composted. The material used is more accurately called mycelium—the body of the fungus—and it comprises fine threadlike

roots that grow underground and mat together naturally. To grow mycelium blocks or panels, a starter base of mycelium is put into a mold along with some agricultural waste, such as corn husks. The mycelium feeds on the cellulose in the husks and grows, and the mold dictates the shape it forms. When it's the desired size and shape, the new block is heated or "cooked off"—so that the mycelium stops growing—and after that, it's ready to use.

BUILDING BIGGER

To produce heavier blocks, experiments are being conducted using mycelium grown onto a heavier scaffolding material. The lightweight versions are already ideal for temporary structures, such as conference halls and pavilions, which traditionally cost a lot and generate more waste; in the longer term, the aim is to produce weightier bricks that can be used for permanent buildings.

MUSHROOM FASHION

Mycelium "leather" is already in production. The method isn't very different from that used to produce mushroom panels. Grown in trays under laboratory conditions and fed a cellulose-rich diet, the fibers eventually form a really dense, thick layer which, like leather, has a strong, uniform consistency, and which can then be sliced and shaped to be used as a leather substitute. The cut slices can then be "tanned" in the way that leather is, using a vegetable process that leaves out the environmentally unfriendly chemicals that are a part of traditional tanning. The finished product has the durability and texture of leather, but without any of the downsides in terms of animal welfare or negative environmental effects—and it's tough enough to be used for anything from shoes to handbags.

104 HOW CAN BACTERIA EAT PLASTIC?

Plastic detritus has made its way all over the world. There are plastic bags in the Mariana Trench, 36,000 feet down in the ocean, and plastic fragments have made their way up almost to the summit of Everest. Finding environmentally friendly ways to dispose of plastic has become an ecological emergency, so it's good news that there may be a natural solution.

MIXING IT UP

Plastic has always been tough to recycle. The very qualities that make it so reliable and durable in use make it a challenge to break down, particularly since the majority of products use a mixture of different plastics rather than just one type. A study from 2017 showed that well under 20 percent of plastic is currently recycled, while production is still rising steeply.

It was originally a chance discovery that the enzymes found in some specific microbes could be used to digest plastic. Enzymes are natural proteins that relish tough digestive challenges: they're responsible for getting rid of plenty of the detritus of the natural world. Not only that, but they have proved capable of discrimination, meaning that they can deal with one element in a mixed plastic product, isolating other elements that may remain useful.

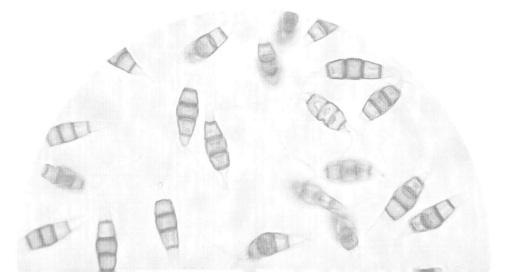

DIGESTING ENZYMES

In 2012, a plant-hunting expedition in Ecuador found a fungus, *Pestalotiopsis microspora*, which contained enzymes that could digest plastic. This was encouraging, but *P. microspora* proved to be slow-acting and quite demanding to maintain. Four years later, *Ideonella sakaiensis* was found in garbage outside a Japanese recycling plant. It represented a breakthrough; it's a bacterium containing enzymes that can be used to break down polyethylene terephthalate, the ubiquitous material used to make plastic bottles, known as PET for short. It was undemanding to keep, proved capable of being tweaked in the lab, and—in tests—proved to work very fast indeed.

Now that there's concrete proof that enzymes can deliver, numerous companies from all over the world are sending scientists to search garbage dumps and landfill sites, keen to find the next as-yet-undiscovered enzyme with a big appetite for plastic.

BREAKING CHEMICAL BONDS

In order to "digest" plastic, enzymes need to break the bonds between the molecules it's made from. This is easier for them to manage in some plastics than others. The material that plastic bottles are made from, PET, for example, consists of conjoined carbon-oxygen molecules, a bond that is also found in tough plant fibers. This is relatively straightforward for enzymes to tackle because they've already had plenty of practice out in the natural world. Some other kinds of plastic, such as polypropylene (which is used for a lot of food packaging, including microwave containers) and polyethylene (which is the material used for most everyday plastic bags), are made from bonded carbon-carbon molecules, which are proving to be a more challenging option to break down.

105 WILL JOBS DIE OUT?

Where job elimination is concerned, humans have been crying wolf for a long time. Back in the 1930s, the economist John Maynard Keynes confidently predicted that people would be working a mere fifteen hours a week within two generations, and forecasts of the death of work have been coming thick and fast ever since.

Aldous Huxley also believed that the working world would become more . . . well, leisurely. In a special future-predicting edition of *Redbook* in 1950, he wrote that workers could expect that everything was due to improve in the near future, including shorter hours, more kindly management,

and substantially higher pay. More than two decades into the twenty-first century, few of us are working less (many might complain they're actually working more). So is there any truth in the idea that jobs will die out?

WORKING 9 TO 5

Automation has been perceived as a threat to jobs since before the Industrial Revolution, but in the past, while new machines may change the nature of a job, they haven't usually eliminated it altogether. The textile workers who destroyed the machines they feared would steal their jobs gave their name to the Luddite fallacy—the belief, which so far has proved to be false, that machines replace people and hugely reduce the job pool. The reality proved to be less straightforward: as mechanization comes, it has disrupted the status quo, but new jobs have tended to arrive with it. The typing pool may be obsolete, for example, but there is now plenty of work for data processors, and while many stores are closing, online retail is booming, moving jobs from one sector to the other.

RISK ASSESSMENTS

However, computerization is changing things much faster today than at any previous time. A large study done by Oxford University in 2013 took a more current view of the job-replacement situation. It looked at over 700 jobs, examining the probability that machines are either already able, or will before long be able, to take over the human role in each. Some aspects of their findings were unsurprising: algorithms can most easily replace jobs that involve repetitive tasks, even those that need a fine degree of precision. They're less successful when a large number of unpredictable variables are introduced into the picture, but they're improving all the time, and may prove more of a threat to the number of jobs overall than they have in the past.

JOB SECURITY

The toughest jobs for a computer to do are at the creative end of the spectrum. Computers are, so far at least, famously unable to either tell or react to jokes reliably, and they can't yet manage the finer points of human-to-human contact. They can't replace artists or courtroom lawyers—a computer can copy a painting, but it can't build a portfolio of unique works; it can analyze a vast quantity of legal documents at speed, but it can't deliver a persuasive argument in front of a jury. If you want an assured job for life, head either for the studio or the courtroom. Or give stand-up comedy a try.

106 HOW SOON WILL WE BE ABLE TO LIVE ON OTHER PLANETS?

We're not yet an interplanetary species; how realistic is it to imagine that humans will be setting up house on other planets anytime soon? The answer depends on who you listen to: some scientists believe that we'll get there (and, given the state of the home planet, that we'll need to get there quite quickly); others think that the space commute will remain in the realms of science fiction for the foreseeable future.

WHICH PLANET?

Is it easier to find ways to travel farther to a very distant planet that has conditions more favorable to life, or to come up with enough adaptations to make life on a closer, less friendly planet possible?

Planet Earth is in the so-called Goldilocks zone—just the right distance from the Sun, neither too hot nor too cold, and with an abundance of water. Our nearest neighbors—the Moon, Mars, and Venus—are far from just right. The Moon is cold, with month-long nights, and lacks atmosphere, in every sense, so radiation would be a problem. And although there is water on the Moon, it's in very short supply—the soil of the Sahara Desert has a hundred times as much. Venus is out of the question: it's boiling hot, each Venusian day lasts 120 Earth days, and its atmospheric pressure is ninety-five times that of Earth's.

WHERE ELSE?

Of the three "neighbor" options, Mars is the best of a bad bunch. It has lighter gravity than Earth and a much thinner atmosphere, and it's considerably colder, though not unsurvivably so. It's believed that there is water under the surface, and it gets sufficient light to make energy-collecting solar panels a possibility. All these factors make it comparatively inviting.

Looking a bit farther afield, Europa, one of Jupiter's moons, is believed to have water, and possibly a more hospitable habitat than Earth's other local planets. Going (much) farther out, the options remain thin. The most hopeful-sounding planet, described as "potentially habitable," is Kepler 442-b, which is large (twice the mass of Earth), and which was discovered by the Kepler telescope in 2015. However, its pluses are set against the very considerable disadvantage that it's located 1,120 light-years away.

COLONIZING MARS

NASA has plans to get a manned mission to Mars in 2030, and a selection of Earth's billionaires are duking it out to join manned space missions, turning space colonization into a competitive activity. How practical it is to speed things up depends on which expert you ask. Michel Mayor, a Nobel Prize–winning astrophysicist at the University of Geneva, doesn't think space colonies will be up and running anytime soon, if at all. "As a species," he points out, "we struggle to reach the Moon." Professor Serkam Saydam, from the University of New South Wales, doesn't agree: "I believe a colony on Mars is going to happen, but between 2040 and 2050 … What I think will happen is that first of all we will … have a colony [on the Moon]. Then we can use the Moon as a [gas] station to get to Mars and beyond. But before 2050 I think we will have settlements on both the Moon and Mars." Watch this space.

107 ARE INSECTS GOING TO BECOME PART OF EVERYONE'S DIET?

By 2050, it's expected that the population of the planet will have reached 9 billion, and questions are already being raised about how we are to be fed. One answer may be entomophagy—eating insects.

BUG IDEAS

If you don't like the idea of chowing down on grasshoppers, termites, or beetles, consider the advantages. Many insects are relatively easy to farm; they tend to do well in hotter climates, and they need much less water than either grazing animals or arable crops. In fact, it costs one tenth of the energy that would be needed to raise 2 pounds of beef—food, water, and land needs—to farm 2 pounds of insect protein. You can either eat insects (and farm fewer of the larger species such as chickens, sheep, or pigs) or you can use them as animal feed (and save on the high cost—both financial and environmental—of animal feed crops).

They have plenty of nutritional advantages, too. Many are high in minerals, proteins, and fatty acids—some species of termites contain as much as 65 percent protein. And although you may not be eating insects yet, plenty of other people are; statisticians tell us that insects form part of the diet of 2 billion people. Africa is the continent with the strongest entomophagic tradition, and around 500 species are eaten, across a range of groups including termites and cicadas, beetles, grasshoppers, and crickets. Many species are also enjoyed in Southeast Asia, Japan, and the Pacific islands.

OVERCOMING THE IMAGE PROBLEM

Given all the benefits, why don't more Westerners eat insects? It seems to be because they don't like the idea of them—rather than the flavor. Back in 1885, in his popular book *Why Not Eat Insects?*, Vincent Holt, a passionate early advocate, wrote: "My insects are all vegetable feeders, clean, palatable, wholesome . . ." before launching into specifics about a whole range of delicacies. Woodlice were especially favored: "I have eaten these, and found that, when chewed, a flavor is developed remarkably akin to that so much appreciated in their sea cousins. Wood-louse sauce is equal, if not distinctly superior, to shrimp."

If you find the names too off-putting for you even to sample the taste, today's marketers are already hard at work fixing some of the terminology in anticipation of an insect-rich future. Following in Holt's footsteps, both woodlice and locusts have appeared in products labeled as "land shrimp," while waxworms, the larvae of the wax moth, have been rechristened "honey bugs" (although enthusiasts claim that their flavor is more like that of pine nuts). In insect-eating societies, crickets have long been ground down to a high-protein flourlike powder used in cooking; in the West, it's being marketed as an energy food.

MINI FARMS

There's a growing, though still niche, trend for home-farmed insects. So-called mini livestock farms may, in time, join other popular small-farm enterprises such as beekeeping and worm composting.

108 IF SOME SPECIES ARE EXTINCT BEFORE THEY'VE BEEN IDENTIFIED, HOW DO WE KNOW THEY WERE EVEN THERE?

Extinction is difficult to declare; in the past, when travel to remote places meant it was tougher to keep accurate records, any species that hadn't been seen for fifty years was declared extinct, with the result that occasionally species believed to be long lost, such as the New Zealand storm petrel, would once more appear on a naturalist's radar.

AN ENDANGERED ABUNDANCE

The International Union for the Conservation of Nature (IUCN) keeps the global tally of the endangered species we know about; its Red List, started in 1964, which has established the population status of more than 138,000 species, gives the number at imminent threat of extinction as 38,500.

When it comes to tiny creatures living in highly specific local habitats in remote places, however, there's a problem.

Faced with the difficulty of knowing what we've got, especially when it comes to insects, the only way to establish the number of species, such as Panamanian golden frogs, going extinct is to make an informed guesstimate. This means taking the species we know about, and the established extinction figures, then taking account of the current threats to biodiversity, from habitat loss to pollution, and estimating both a global species count and a count of what proportion of that faces extinction. In 2019 an extensive and sobering study carried out by the United Nations concluded that of an estimated 8.1 million species in the world, around a million are threatened with imminent extinction.

QUIZ
THE FUTURE

Sometimes the future is impossible to predict. But sometimes you can know what to expect. Take this quiz to find out how much you know about what we might know!

QUESTIONS:

1. What might be the best careers to choose if you want your field of work to last?

2. Bricks made from fungi have environmental advantages; what are they?

3. For the future, which destination looks most habitable: the Moon, Mars, or Venus?

4. What was the Luddite fallacy?

5. If you practice entomophagy, what do you do?

6. What is spectral cloaking?

7. What do the initials IUCN stand for?

8. Which species have been rechristened "land shrimp" and "honey bugs" by food marketers?

9. Why might the *Ouroboros Steak* project have been considered vegetarian?

10. What is the so-called Goldilocks zone?

Turn to page 249 for the answers.

QUIZ ANSWERS

COSMOLOGY

1. Up to 11 miles

2. Gravity pulls them together

3. The Inouye Solar Telescope in Hawaii gives the most detailed pictures of the Sun's surface

4. It slows down

5. Around 3.8 billion years old

6. At a well-chilled 0.00036 degrees kelvin, it was a tiny piece of aluminum, in a test carried out in Boulder, Colorado

7. The scientist Fritz Zwicky, in 1933

8. A galaxy can be up to 300,000 light-years (that's 1.7 quintillion miles) across

9. Because light pollution from Los Angeles spoiled its view

10. It's the stretching process that would be induced by gravity if you were to fall into a black hole

ASTRONOMY

1. It's an event in which the Sun, Moon, and Earth align

2. A lunar eclipse

3. Gravitational forces mean that the Earth's width varies by up to 13 miles

4. It would protect you up to around 250°F

5. Around 370,000 miles

6. It was first recorded in ancient China, in 240 BCE

7. "Hoba," which weighs 66 tons and is 10 feet wide

8. No, because the Earth's rotational axis moves; in the remote future, other stars will take a turn as the North Star

9. Not until 2492

10. Two, although one of them is expected to collide with Mars at some point in the next 10 million years

EARTH MATTERS

1. In the Northwest Pacific

2. It's the measure by which the power of a hurricane is measured

3. 11 years

4. In El Salvador, on the slopes of the Izalco Volcano

5. An area in the North Atlantic, where melting glacier ice is adding additional chill

6. Approximately 70 percent

7. It describes what happens when two hurricanes collide in midair

8. Between 500 and 1,000 years, depending on the climate around it

9. More than a hundred new minerals are currently being identified every year

10. 2008's Storm Marco, which was only around 12 miles across

FORCES AND MATERIALS

1. It's currently helium, which melts at -458°F

2. It's the process whereby the atoms in a material decay first by half, and then by half that half, ad infinitum

3. It only exists in scientists' imaginations at the moment, but a ray cat would be a cat designed to emit a warning glow

4. Static is caused by the buildup of an electrical charge in one place

5. You would get heavier and shorter— and time would seem to slow down around you

6. Every magnet has both a positive (north) and a negative (south) side, and however many parts you cut a magnet into, each piece will still maintain those sides

7. It states that if you want something to move, it will take energy to do it

8. In principle you could manufacture a big diamond, in practice it would be prohibitively costly to do so

9. Settling atoms, forming crystalline structures, make it expand

10. At Incheon Airport in South Korea

LIFE ON EARTH

1. Yes, *T. rex* had 60, nearly double the human count of 32

2. Five so far; many scientists believe that we are already in the middle of the sixth

3. Arctic worms are a warning sign that a rare habitat is being warmed and eroded

4. Living specimens were discovered when it was believed to have been extinct for 65 million years

5. 12

6. *Turritopsis dohrnii*, the immortal jellyfish, can grow younger as well as older

7. In 2012, a study determined that it would take 6.8 million years to destroy DNA completely

8. It's used for detecting modern art fakes which are being sold as masterpieces

9. 98.8 percent

10. It's a global list of the species known to be at risk from extinction

BOTANY

1. Fossil trees have been found that date from the late Devonian era, 385 million years ago

2. 56, by a clover breeding specialist in Japan

3. Its flower warms to around 98° F (36.7°C), matching the temperature of a rotting animal

4. That of *Rafflesia arnoldii*, the stinking corpse lily; it weighs around 24 pounds

5. Up to three decades

6. Groups of trees that share an interconnected root system

7. It prevents too much water from being lost

8. The process by which a plant "deceives" an insect into pollinating its flowers

9. They're extremely fine-barbed filaments that some plants release when threatened

10. They produce toxins that they spread into the mycorrhizal network to ensure other trees don't intrude on their space

ZOOLOGY

1. It can change texture in a moment, going from thick and gluey to thin and runny, and back again

2. An estimated 1.7 million

3. Because with water to carry its weight, it's less controlled by gravity than if it lived on land

4. 35 percent, 14 percent higher than the oxygen level today

5. Because the bulbs carry the possibility of life, as they can grow a new plant

6. Their bodies naturally compensate for the lower levels of oxygen available

7. They can produce clicking sounds that jam the bat's echolocation system

8. It's the highly modified left front tooth (the right front tooth is normal size)

9. The fastest dive recorded hit a rate of 240 mph

10. Chickens—there are an estimated 19 billion on the planet, compared to 7.8 billion humans

THE HUMAN BODY

1. On the top of its head; it's the soft spot where the plates of the skull haven't yet fused together

2. 5 feet, 6 inches—taller than average for his times

3. Between 2 and 4 pounds

4. In your brain; it controls your emotional responses and how you move your face expressing them

5. The didgeridoo

6. Chemicals, including the atmospherically named putrescine and cadaverine, which are produced by the breakdown of amino acids in a dead body

7. Special white blood cells called phagocytes absorb the dead cells and recycle any useful ingredients

8. No; there have been periods in the past where people have become shorter, and others during which they've grown taller

9. An estimated 50 billion cells per person die every day

10. Yes, there are two ways: knismesis is a "light touch" tickle, while gargalesis is more heavy-handed

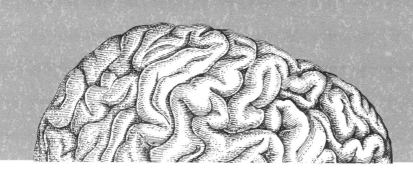

THE HUMAN BRAIN

1. Up to half, but no more than that—and it would need to be removed by an expert

2. The hypothalamus, at the base of your brain

3. A rare condition in which the subject remembers every single daily event of their life in minute detail

4. Two-thirds

5. She believed that she had swallowed a grand piano made from glass

6. You'd find it difficult to digest fatty foods

7. The adrenal gland

8. Because the humidity in the air means that sweat won't evaporate from your skin

9. It's the level and degree of stress that stimulates you to make an effort (unlike chronic stress, which can drag you down)

10. Not necessarily; there are variants in which you might be able to experience sounds as smells, or to "taste" words

MEDICINE

1. Karl Landsteiner

2. In the *Sushruta Samhita*, an encyclopedia of health written in the sixth century BCE

3. In 1978, as the result of a laboratory accident; its global eradication was officially announced by the WHO in 1980

4. They are infections that have developed multiple resistances to a wide range of antibiotics

5. They're different terms for the same operation: making a hole in the skull to release pressure

6. You'd be kept at −320°F

7. The preservation of a body by immersing it in honey

8. You can give blood to all other groups, although if you need a transfusion yourself you can only have one from a fellow HH grouper

9. Around 60 percent died

10. The Karnavedha

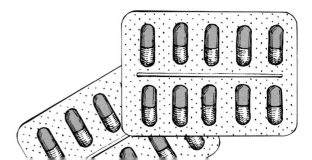

COMPUTING AND TECHNOLOGY

1. Heat-assisted magnetic recording

2. They're naturally occurring structures in metals, which, in a computer, act as tiny magnets

3. Just three: red, green, and blue are combined in different amounts

4. Three: incoming primary wire, outgoing secondary wire, and a magnetic core

5. In terms of environmental impact, it's around 50 percent cleaner

6. It is the information-storing unit of a quantum computer

7. They can carry ten times as much information, over a much greater distance, and with greater security

8. Because the radiation from the signals it puts out is non-ionizing, and thus safe

9. At the end of 2020, it was calculated that there were over 10 million on the roads

10. It forces two hydrogen atoms to fuse together into a single helium atom, a process that releases 10 million times as much energy per pound as fossil fuels do

THE FUTURE

1. Creative art, courtroom law, and stand-up comedy are all fields that, long term, seem likely to survive automation

2. They're carbon-neutral, fire-resistant, and compostable

3. Mars

4. The belief that machines replace people and deprive them of jobs

5. You eat insects

6. A developing technique for rendering objects invisible

7. International Union for the Conservation of Nature

8. Woodlice and locusts, and wax worms

9. Because it allows people to grow edible meat from their own cells

10. It's the area in space occupied by Earth, at just the right distance from the Sun to encourage life

INDEX

CREDITS

1, 3, 10, 29, 49, 67, 87, 105, 125, 145, 163, 185, 205, 223 © Naci Yavus | Shutterstock • 1,3, 10, 30, 31, 56, 68, 69, 84, 98, 107, 146, 147, 246, 253 © Morphart Creation | Shutterstock • 6, 88, © Warpaint | Shutterstock • 7, 21, 31 © Nerthuz | Shutterstock • 7, 8, 10, 32, 33, 44, 86, 250, 254 Arthur Balitskii © | Shutterstock • 8, 99 © Blackboard | Shutterstock • 8, 145, 150, © Lifeking | Shutterstock • 8 © Champ Nattapon | Shutterstock • 8 © Ekaterina Glazkova | Shutterstock • 9 © Ezepov Dimitry | Shutterstock • 9, 28 pikepicture © | Shutterstock • 9 © Captureandcompose | Shutterstock • 13 © ixpert | Shutterstock • 13, 16, 34, 35, 62, 63, 95, 106, 126, 133, 141, 149, 162, 165, 168, 188, 190, 198, 208, 220, 234, 237, 243 Wikimedia.org • 14 © local_doctor | Shutterstock • 15 © Antares-StarExplorer | Shutterstock • 16, 45 wikimedia • 17 © ex_artist | Shutterstock • 20 © inrainbows | Shutterstock • 23, 105, 122 © Bodor Tivadar | Shutterstock • 25 © sakkmesterke | Shutterstock • 28, 38 d1sk | Shutterstock • 33 © Dotted Yeti | Shutterstock • 34, 251 © nickolai_self_taught | Shutterstock • 36 © Mary Frost | Shutterstock • 39 © Michael Rosskothen | Shutterstock • 40 © Good Studio | Shutterstock • 41 © Vladi333 | Shutterstock • 42 © WWWoronin | Shutterstock • 43 © Hoika Mikhail | Shutterstock • 45 NASA, Wikimedia.org • 46 © ArtMari | Shutterstock • 49 © Fotokina | Shutterstock • 50, 255 © Nadya Dobruynina, alexblacksea, La puma | Shutterstock, • 52 © sar14ev | Shutterstock • 53 © SvetlanaARTdreams | Shutterstock • 54 © amenic181 | Shutterstock • 55 © akr11_ss | Shutterstock • 57 © Kseniakrop | Shutterstock • 60 © Rashevskyi Viacheslav | Shutterstock • 64, 244 © Suns07butterfly | Shutterstock • 66, 68 © | Shutterstock • 70, 74 © Everett Collection | Shutterstock • 72 © Daniel Prudeck | Shutterstock • 73 © Janista | Shutterstock • 78 © rook76 | Shutterstock • 76 pixabay.com • 79, 116, 158, 176 Getty Images • 80 © Mark S Johnson | Shutterstock • 81 © topseller | Shutterstock • 82, 252 © Alexander P | Shutterstock • 86, 90, 97, 100, 246 © Eric Isselee | Shutterstock, Liliya Butenko • 88 Paul C. Sereno, Wikimedia.org • 89, 154 Alamy • 94 © Sergey Mikhaylov | Shutterstock • 101 © Prokhorovich | Shutterstock • 104, 111, 119 © Le Do | Shutterstock, Sap Green Illustration • 107 © lamnee | Shutterstock • 108, 169 © Maisei Raman, o-sun | Shutterstock • 109, 110 © le Do | Shutterstock • 111 © tanatat | Shutterstock • 114 © Troutnut | Shutterstock • 115 © graphuvarov | Shutterstock • 117 plantillustration.org • 118 © Heather Lucia Snow | Shutterstock • 120 © Nella | Shutterstock • 121 © Yulick_art | Shutterstock • 122 © | Shutterstock • 124, 126, 130, 132 © Hein Nouwens, Serafima Antipova | Shutterstock • 125, 136, 247 © Alex Rockheart | Shutterstock • 129, 130

© Andrey oleynik, Christoph Burgstedt | Shutterstock • 130 © Elizaveta Meletyeva | Shutterstock • 131 Stocksnapper © | Shutterstock • 133 © Vector Tradition | Shutterstock • 134 © photomaster | Shutterstock • 138 © AVA Bitter | Shutterstock • 139 © Menno Schaefer | Shutterstock • 140 © Cimmerian | Shutterstock • 142 © Natalie Jean | Shutterstock • 144, 147, 188, 189, 196, 197 wellcomecollection.org • 146 Pixologicstudio Science Photo Library • 148 Sebastian Kaulitzki, Shutterstock 152 © Anna Kireiva | Shutterstock • 153 © Varlamova Lydmila | Shutterstock • 155 © Nathapol Kongseang, Arkhrakrit | Shutterstock • 157 © RetroClipArt | Shutterstock • 159 © pavlematic | Shutterstock • 160 © Crystal Eye Studio | Shutterstock • 162, 182 © Jolygon | Shutterstock • 163, 172 © omyim1637| Shutterstock • 164 © Hurst Photo | Shutterstock • 166 World History Archive, Alamy • 167, 222, 227 © Epine | Shutterstock • 168 Kindlena, Shutterstock • 170 © John_Dakapu | Shutterstock • 173 © Lisa F Young | Shutterstock • 174, 175, 248 © Hein Nouwens | Shutterstock • 177 © Iaroslav Neliubov | Shutterstock • 179, 207 © Everett Collection | Shutterstock • 180 © Jolygon | Shutterstock • 181 © evenfh | Shutterstock • 184, 186, 248 © Alisa pravotorova | Shutterstock • 184, 201 © Neveshkin Nikolay | Shutterstock • 187 © Kateryna Kon | Shutterstock • 191 © Aurelija Diliute | Shutterstock • 192 © Channarong Pherngjanda | Shutterstock • 193 © akr11_ss | Shutterstock • 194 © Emmily | Shutterstock • 198 G. Elliot Smith, Wikimedia.org • 199 © Morkhatenok | Shutterstock • 200 © oksana2010 | Shutterstock • 202 © Fabio Pagani , Koya979 | Shutterstock • 204, 219, 249 © ZubetroN | Shutterstock • 204, 206 © mkos83 | Shutterstock • 204, 219, 249 plantillustrations.org • 205, 209 © cigdem | Shutterstock • 205, 210 © Champnatatoon | Shutterstock • 212 © Nordroden | Shutterstock • 213 © Ink Drop| Shutterstock • 214 © Sstranger | Shutterstock • 215 © Makhnach_S | Shutterstock • 216 © Martial Red | Shutterstock • 217 © Armin Van | Shutterstock • 218 © Yravetor | Shutterstock • 219 © dwersteg | Shutterstock • 222 © SpicyTruffel | Shutterstock • 222, 230 © Komleva | Shutterstock • 222, 232 Trevor Patt • 225 © Andrey tiyk | Shutterstock • 226 © nevodka | Shutterstock • 228 © Uncle Leo , Oleg Vinnichenko | Shutterstock • 229 © DnBr | Shutterstock • 230 © ktsdesign | Shutterstock • 233 © Yulia Panova| Shutterstock • 234 Matthew Schink, Wikimedia.org • 235 © Larina Marina | Shutterstock • 236 © chippix | Shutterstock • 238 © Victoria Sergeeva | Shutterstock • 239 © Suparmotion | Shutterstock • 240 © nicemyphoto | Shutterstock • 241 © Romas_Photo | Shutterstock • 242 Brian Gratwicke • 244 © Angioweb | Shutterstock